ENGINEERING TOOLS, TECHNIQUES AND TABLES

COMPUTATIONAL ENGINEERING

DESIGN, DEVELOPMENT AND APPLICATIONS

ENGINEERING TOOLS, TECHNIQUES AND TABLES

Additional books in this series can be found on Nova's website under the Series tab.

Additional E-books in this series can be found on Nova's website under the E-book tab.

COMPUTER SCIENCE, TECHNOLOGY AND APPLICATIONS

Additional books in this series can be found on Nova's website under the Series tab.

Additional E-books in this series can be found on Nova's website under the E-book tab.

ENGINEERING TOOLS, TECHNIQUES AND TABLES

COMPUTATIONAL ENGINEERING

DESIGN, DEVELOPMENT AND APPLICATIONS

JACLYN E. BROWNING
AND
ALEXANDER K. MCMANN
EDITORS

Nova Science Publishers, Inc.
New York

For permission to use material from this book please contact us:
Telephone 631-231-7269; Fax 631-231-8175
Web Site: http://www.novapublishers.com

LIBRARY OF CONGRESS CATALOGING-IN-PUBLICATION DATA

ISBN 978-1-61122-806-9

Published by Nova Science Publishers, Inc. † New York

CONTENTS

PREFACE

Computational engineering encompasses both development and application of software tools and systems, including physical, mathematical, and geometric modeling, solution algorithms, computer simulations, and visualization, analysis, interpretation, synthesis and the use of computed results to solve practical problems. This book presents current research in the study of computational engineering, including computer-aided design of tissue engineering applications; crash test simulations and their application in automotive design and quantum neural computation.

Tissue engineering (TE) is an emerging multidisciplinary field that draws on expertise from medicine, biology, chemistry, genetics, engineering, computer and life science. Its mission is to discover solutions to one of the most challenging medical problems faced by humans: replace tissue and organ functions when damage is beyond natural recovery process. A precondition for successful TE is having an adequate understanding of the principles of tissue genesis. The goal is to apply that knowledge to produce functional tissue replacements suitable for clinical use. Specifically achieve biological-inspired, biocompatible, tissue-mimetic structures that, when implanted *in vivo*, restore or improve failed or compromised human tissue and/or organ function. Impressive progress in human tissue regeneration followed development and implementation of advanced technologies that enabled better understanding and control signalling within microenvironments during growth and maturation of tissue functionalization. In particular, the latest generation of bioreactors have demonstrably improved *in vitro* tissue maturation prior to implantation. That achievement was made possible by two engineering advances: i) repeatable and automated bioprocesses, and ii) recapitulation of key physiologic, physicochemical and mechanical cues *in vitro*. Despite this progress critical, large gaps in our knowledge are slowing progress. For example, how can cell level operating principles and environmental cues be orchestrated in advanced bioreactors to enable the formation of a physiological-like functional tissue? What are those cell level operating principles? Because the tissue is developing *ex vivo*, will the orchestration need to be different in important ways from that occurring during organogenesis? When detailed information is limited, uncertainties are large, and feasible wet-lab experiments are limited by costs and other factors, *in silico* exploratory modelling and simulation can be a cost-effective adjunct strategy for answering those and related questions. It can help achieve critically needed mechanistic insight into complexities of suitable cellular responses to the conditioning procedures. Improving and experimenting upon multi-attribute, multiscale computational models is a relatively new, scientific approach for i) evaluating cellular responses to the conditioning procedures within bioreactors, ii) predicting or anticipating the dynamic modification of the cellular behaviours during culture as a function of external stimuli, and iii) discovering relevant features and protocols for optimizing bioreactor working conditions (i.e., the external stimuli for the cells), in terms of interactions between cells and the bioartificial hosting environment. Within Chapter 1 the authors discuss how coupling i) computational fluid dynamics (CFD) and ii) multi agent systems (MAS) modelling methods is enabling rationale design and subsequent establishment of *in silico bioreactors*. Problems associated with designing TE experiments can be explored using *in silico* high-throughput experiments, and plausible solutions can be identified in advance by creating a software framework, which incorporates a variety of phenomena known to influence tissue growth, along with a model of cell population dynamics. Upon maturation, the approach is expected to provide exploitable insight into how tissue generation and maturation emerge and can be controlled. That insight can be leveraged to fine-tune system parameters to achieve desired cellular responses during *in vitro* conditioning while reducing reliance on costly physical experiments. Computational modelling and simulation in TE is presented as a diverse, active, and powerful trans-disciplinary expansion of scientific and engineering methods for overcoming many of the current limitations in identifying optimal regenerative therapies for diseased and damaged tissues.

Crash test simulations possess a stochastic component, related with physical and numerical instabilities of the underlying crash model and uncertainties of its control parameters.

Scatter analysis is a characterization of this component, usually in terms of scatter amplitude distributed on the model's geometry and evolving in time. This allows to identify parts of the model, possessing large scatter and most non-deterministic behavior. The purpose is to estimate reliability of numerical results and predict corresponding tolerances.

Causal analysis serves a determination of cause-effect relationships between events. In context of crash test analysis, this usually means identification of events or properties causing the scatter of the results. This allows to find sources of physical or numerical instabilities of the system and helps to reduce or completely eliminate them.

Sensitivity analysis considers dependence of numerical results on variations of control parameters. It allows for identification of the parameters with the largest influence to the results and the parts (locations) of the model where the impact of such variations is considerable.

Metamodeling of simulation results allows the authors to make a prediction of the model behavior at new values of control parameters, for which a simulation has not been performed yet. These values of parameters are usually intermediate with respect to existing ones, for which simulations are available. Predictions are needed for analysis and optimization of product properties during the design stage and are usually completed with a control simulation performed at the optimal parameter values.

In Chapter 2, statistical methods which allow for an efficient solution of these tasks are reviewed, and a novel method is introduced and its effiency demonstrated for benchmark cases.

Classical computing systems perform classical computations (i.e., Boolean operations, such as AND, OR, NOT gates) using devices that can be described classically (e.g., MOSFETs). On the other hand, quantum computing systems perform classical computations using quantum devices (quantum dots), that is, devices that can be described only using quantum mechanics. Any information transfer between such computing systems involves a state measurement. Chapter 3 describes this information transfer at the edge of topological chaos, where mysterious quantum-mechanical linearity meets even more mysterious brain's nonlinear topological complexity, in order to perform a super-high-speed and error-free computations.

The motivation behind mathematically modeling the *human operator* is to help explain the response characteristics of the complex dynamical system including the human manual controller. In Chapter 4, the authors present two approaches to human operator modeling: classical linear control approach and modern nonlinear control approach. The latter one is formalized using both fixed and adaptive Lie-Derivative based controllers.

A global secant relaxation(GSR)-based accelerated iteration scheme can be used to carry out the incremental/iterative solution of various nonlinear finite element problems. This computation procedure can overcome the possible deficiency of numerical instability caused by local failure existing in the iterative computation. Moreover, this method can efficiently accelerate the convergency of the iterative computation. This incremental/iterative analysis can consistently be carried out to update the response history up to a near ultimate load stage, which is important for investigating the global failure behaviour of a structure under certain external cause, if the constant stiffness is used. Consequently, this method can widely be used to solve general nonlinear problems.

In Chapter 5, mathematical procedures of Newton-Raphson techniques in finite element methods for nonlinear finite element problems are summarized. These techniques are the Newton-Raphson method, quasi-Newton methods, modified Newton-Raphson methods and accelerated modified Newton-Raphson methods.

Numerical results obtained by using various accelerated modified Newton-Raphson methods are used to study the convergency performances of these techniques for material nonlinearity problems and deformation nonlinearity problems, separately.

In: Computational Engineering
Editors: J. E. Browning and A. K. McMann

ISBN: 978-1-61122-806-9

Chapter 1

IN SILICO BIOREACTORS FOR THE COMPUTER-AIDED DESIGN OF TISSUE ENGINEERING APPLICATIONS

F. Consolo[1,3,1], L. Gaetano[1,2], G. Di Benedetto[1], C. A. Hunt[2], F. M. Montevecchi[1] and U. Morbiducci[1]
[1] Department of Mechanics, Politecnico di Torino, Torino, Italy
[2] Department of Bioengineering and Therapeutic Sciences, University of California, San Francisco, CA, US
[3] Department of Bioengineering, Politecnico di Milano, Milano, Italy

ABSTRACT

Tissue engineering (TE) is an emerging multidisciplinary field that draws on expertise from medicine, biology, chemistry, genetics, engineering, computer and life science. Its mission is to discover solutions to one of the most challenging medical problems faced by humans: replace tissue and organ functions when damage is beyond natural recovery process. A precondition for successful TE is having an adequate understanding of the principles of tissue genesis. The goal is to apply that knowledge to produce functional tissue replacements suitable for clinical use. Specifically achieve biological-inspired, biocompatible, tissue-mimetic structures that, when implanted *in vivo*, restore or improve failed or compromised human tissue and/or organ function. Impressive progress in human tissue regeneration followed development and implementation of advanced technologies that enabled better understanding and control signalling within microenvironments during growth and maturation of tissue functionalization. In particular, the latest generation of bioreactors have demonstrably improved *in vitro* tissue maturation prior to implantation. That achievement was made possible by two engineering advances: i) repeatable and automated bioprocesses, and ii) recapitulation of key physiologic, physicochemical and mechanical cues *in vitro*. Despite this progress critical, large gaps in our knowledge are slowing progress. For example, how can cell level operating principles and environmental cues be orchestrated in advanced bioreactors to enable the formation of a physiological-like functional tissue? What are those cell level operating principles? Because the tissue is developing *ex vivo*, will the orchestration need to be different in important ways from that occurring during organogenesis? When detailed

---

[1] E-mail address: filippo.consolo@polimi.it

information is limited, uncertainties are large, and feasible wet-lab experiments are limited by costs and other factors, *in silico* exploratory modelling and simulation can be a cost-effective adjunct strategy for answering those and related questions. It can help achieve critically needed mechanistic insight into complexities of suitable cellular responses to the conditioning procedures. Improving and experimenting upon multi-attribute, multiscale computational models is a relatively new, scientific approach for i) evaluating cellular responses to the conditioning procedures within bioreactors, ii) predicting or anticipating the dynamic modification of the cellular behaviours during culture as a function of external stimuli, and iii) discovering relevant features and protocols for optimizing bioreactor working conditions (i.e., the external stimuli for the cells), in terms of interactions between cells and the bioartificial hosting environment. Within this chapter we discuss how coupling i) computational fluid dynamics (CFD) and ii) multi agent systems (MAS) modelling methods is enabling rationale design and subsequent establishment of *in silico bioreactors*. Problems associated with designing TE experiments can be explored using *in silico* high-throughput experiments, and plausible solutions can be identified in advance by creating a software framework, which incorporates a variety of phenomena known to influence tissue growth, along with a model of cell population dynamics. Upon maturation, the approach is expected to provide exploitable insight into how tissue generation and maturation emerge and can be controlled. That insight can be leveraged to fine-tune system parameters to achieve desired cellular responses during *in vitro* conditioning while reducing reliance on costly physical experiments. Computational modelling and simulation in TE is presented as a diverse, active, and powerful trans-disciplinary expansion of scientific and engineering methods for overcoming many of the current limitations in identifying optimal regenerative therapies for diseased and damaged tissues.

1. Introduction: Tissue Engineering Paradigm in the Regenerative Medicine Field

Until the twentieth century, medical treatments were rarely capable of reversing pathologic processes leading to loss of tissue function. Surgical treatments were primarily extirpative, providing limited, if any, functional restoration of damaged tissues. As knowledge in pathophysiology has grown, from the cellular to the subcellular until the biomolecular level, dysfunctional genes, molecules, cells and organs have been successfully restored or replaced with a functional duplicate, following a process that is defined as regenerative medicine [Jenkins et al., 2003]. Regenerative medicine (RM) is an emerging multidisciplinary field which, joining together expertise in medicine, biology, chemistry, genetics, engineering, computer science, robotics, and others, aims at finding solutions to some of the most challenging medical problem faced by humans, i.e. the recovery of damaged/injured tissue and organ functions. RM as a new way of treating injuries and diseases originates as a consequence that most human tissues are unable to spontaneously regenerate after severe trauma [Stoltz et al., 2006]. This is why biotherapies as well as the manufacturing of bioartificial or bioinspired tissue replicates represent today a promising alternative in clinics and therapeutics. In the last few years, this innovative approach has rapidly come out as an efficient alternative strategy in the treatment of diseases of several human tissues as bone, cartilage, tendons, ligaments, skin, blood vessels, skeletal and cardiac muscle, heart valves, nerves, bladder and so on. At the beginning of the twenty-first century, RM is building platforms for even more complex organ recovery, as for kidney, liver and pancreas. This goal is being realized by collaborative efforts in life science and bioengineering techniques. There are mainly three concentrations in the field of regenerative medicine, which are (fig. 1): i) transplantation surgery; ii) gene/cell therapy; iii) tissue engineering.

Allogenic organ transplantation represents today the only clinical solution for the treatment of end-stage organ failure as severe cardiac heart, hepatic or kidney failure [Jenkins et al., 2003]. This approach is, however, limited due to the persistent shortage in donors and also because of the natural physiologic response (rejection) of the body to foreign tissues. With genetic or cellular manipulation, cells, tissues and organs may be engineered to more closely represent an autologous status [Jenkins et al., 2003]. Although gene therapy, attempting to directly manipulate the genome of a cell, is for sure the new most promising frontier in RM, investigation of cellular and genetic manipulation techniques goes beyond the aim of the present chapter, which principally focuses on tissue engineering applications. Nevertheless, combination of these approaches may amplify the natural healing process or take over the function of a permanently damaged organ. This is clearly summarized in figure 2, depicting the historical progress and the future perspective of RM.

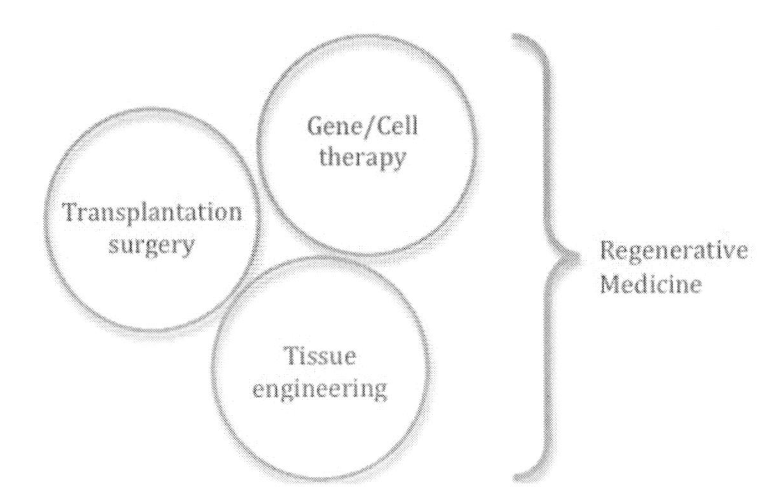

Figure 1. Regenerative medicine approaches can be classified in its three main subcategories, which are: i) transplantation surgery, ii) gene/cell therapy and iii) tissue engineering.

A universally accepted definition of tissue engineering (TE), as stated by Langer and Vacanti, is "an interdisciplinary field that applies the principles of engineering and life sciences toward the development of biological substitutes that restore, maintain, or improve tissue function or a whole organ" [Langer and Vacanti, 1993]. Fundamentally, TE is based on the assumption of "understanding the principles of tissue growth, and applying this to produce functional replacement tissues for clinical use" [MacArthur, 2005]. For more than ten years, both medical practice and health science have taken advantage from progress in the field of fundamental biology, chemistry, physics and ultimately mechanics and their engineering applications. As a consequence, TE has grown up as the combination of these broad set of tools at the interface of the biological, medical and engineering sciences, employing viable cells to *in vitro* promote new tissue-like structure formation, thereby producing therapeutic as well as diagnostic benefit [Stoltz et al., 2006]. The effort of engineering tissue regeneration has been up to now attempted by following different approaches. In the most frequent paradigm, first, viable donor cells are collected and either cultured and expanded, with or without modifying their properties (specific gene modification). After an *in vitro* static cultivation period, which can or cannot include the use of a 3D porous support, called scaffold (i.e., a substrate for cell culture primarily made of synthetic or biologically derived material) the engineered constructs are transferred within a bioreactor where physiological-like physicochemical and mechanical cues can be recapitulated and controlled *in vitro*. Once the tissue has reached fully maturation within bioreactors, the construct is implanted in the appropriate anatomic location, like prosthesis [Rabkin and Schoen, 2002]. In figure 3, a schematic of the scaffold/bioreactor-based TE approach is provided.

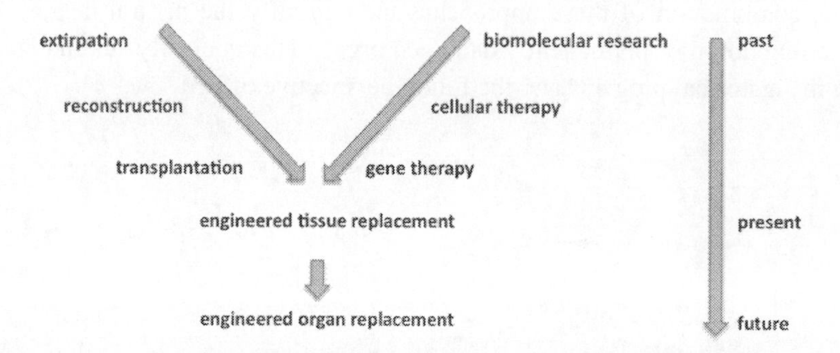

Figure 2. Historical progress and future perspective in the regenerative medicine field.

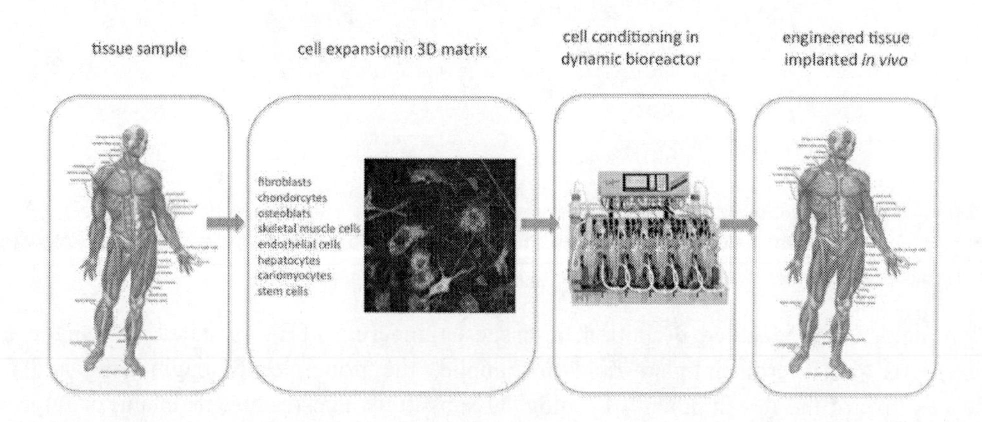

Figure 3. Schematic diagram showing the scaffold-based TE methodology.

The issue of *in vitro* cell culture is multifaceted as it involves i) cell genetics, for the choice of the proper cell source (progenitor, differentiated neonatal/adult or genetically modified cells), ii) cell biochemistry, for the choice of the scaffold material promoting cell-substrate favourable interactions and for the selection of the chemical composition of the cultivation environment (i.e., the medium composition: oxygen tension, nutrient and grow factor concentration, etc.) and iii) cell mechanics, for the design of the appropriate dynamic conditioning protocol within bioreactors. Nevertheless, practical considerations must be added to the previous matters, which are related to the design and optimization of the material, the size, the shape and the morphology and topology of bioengineered scaffolds: all of these i) must ensure biocompatibility, i.e. not inducing the onset of toxic events or producing immune response while appropriately integrating *in vitro* and *in vivo*, ii) must promote cell adhesion and permitting cell growth, proliferation and tissue maturation and functionalization, and iii) must allow the physiological milieu to circulate everywhere during cellular growth, so to provide cells within an environment rich in oxygen and nutrients.

Different experimental procedures exist in TE, which are tissue-specific. For example, the amount of tissue produced by *in vitro* cultures is generally greater when 3D porous supports are used instead of standard monolayer cultures. In such a condition, the construct (i.e., the scaffold-cells assembly) is expected to go through a maturation process, in which cells attach to the substrate surface, proliferate and reorganize mimicking the structure of native tissues while an interlinked extracellular matrix (ECM) is formed and the scaffold degraded. Traditionally, TE has been applied using formulations of cells in conjunction with chemical and mechanical conditioning in the attempt to fabricate biomimetic tissues and organs. Mechanical conditioning has been proven to influence the differentiation, maturation and gene expression of particular cellular phenotypes, as musculoskeletal, cardiac, endothelial or cartilaginous cells. As an example, in engineering cartilage tissue, the "quality" of the *in vitro* engineered cartilage depends on the intensity, magnitude and frequency of the mechanical stress imposed during dynamic conditioning within bioreactors as well on the established 3D environment [Jung et al., 2009]. This finding is coherent with the following observed physiological mechanism: in native tissue a prolonged immobilization, implying absence of mechanical stimulation, results in weakening of cartilage, bone and muscles and to the consequent decrease of their respective body mass and volume. Also, in bone TE, it has been shown how shear stresses can have a major influence on ECM deposition [Botchwey et al., 2003; Pierre and Oddou, 2007; Porter et al., 2007], which is the first step of new bone synthesis. On the other hand, subjecting blood vessel cells to non-physiologic hydrodynamic mechanical stress has been proposed as a key factor regulating and influencing the onset of vascular pathologies [Malek et al., 1999].

With bioreactor cultivation, it is possible to better control and regulate oxygen and nutrient delivery to the viable cells as well as to enhance tissue formation by applying mechanical stimuli to the engineered constructs (scaffold and cells) *in vitro* [Couet and Mantovani, 2010]. Currently, the majority of TE-based clinical applications in human regenerative medicine have been achieved using scaffold-mediated techniques joined with bioreactor-based dynamic tissue conditioning [Gomez, 2007]. Successful results have been reached in growing human skin [Andreadis, 2007] non-load bearing cartilage [Saim, 2000], and human urinary bladders [Atala et al., 2006]. These milestones enforce the idea that the engineering of organs or tissues such as bone, liver, heart valves, myocardium and blood vessels is feasible, within the next few decades.

2. Bioreactor Technology in Tissue Engineering and Regenerative Medicine

During the last decade, noticeable advances have been achieved in human tissue regeneration with the inclusion of advanced technologies that enabled the possibility to better understand

and control signalling within the microenvironments during cell growth and the new-tissue functionalization. In particular, bioreactors have been developed in order to improve the tissue maturation process *in vitro*, prior to the *in vivo* engineered tissue implantation. A universally recognized definition of what a bioreactor is does not exist. Bilodeau and Mantovani proposed that a bioreactor could be defined as "any apparatus that attempts to mimic and reproduce physiological conditions in order to maintain and encourage cell culture for tissue regeneration" [Biolodeau and Mantovani, 2006]. In TE, a "bioreactor" was initially a simple Petri dish undergoing mixing and incubated at a controlled temperature. Over time, the evolution of technical tools and the need of a better comprehension of cellular mechanics have led to the development of more sophisticated *in vitro* dynamic systems [Lyons and Pandit, 2005; Piccinini et al., 2009]. Bioreactors, which are to date a key technology in TE [Shachar and Cohen, 2003], offer attractive advantages: beyond matching the definition of repeatable and automated bioprocesses, they are designed to recapitulate the physiologic physicochemical and mechanical cues *in vitro*. The need for such devices is proven by the statement that, although the ideal *in vitro* conditions for the formation of functional tissue constructs are not known, several studies have shown that physicochemical and mechanical preconditioning benefits the generation of tissue engineered structures [Niklason et al., 1999; Nerem and Seliktar, 2001; Hoerstrup et al., 2000].

To date, bioreactors can be classified as:

- 3D dynamic culture model systems, enabled to outline specific aspects of the actual *in vivo* milieu as well for the investigation of 3D *in vitro* cell growth
- graft manufacturing devices, implementing bioprocesses supporting the production of grafts for pre-clinical and clinical use [Piccinini et al., 2009]
- bioartificial devices utilized as extra-corporeal organ support devices [Miller, 2000]

Currently, manufacturing of TE products requires a great number of manual operations performed by trained personnel, with potential drawbacks such as operator variability, cross-contamination risks, limited scale-up possibilities and high manufacturing costs. Hence, the shortening and simplification of procedures represent solution strategies for a safer and wider implementation of TE in clinical practice. Bioreactors allow aseptic feeding and sampling to follow tissue development, maximizing the use of automated processing steps to increase reproducibility. Following this perspective, bioreactors represent the key element for the development of automated, standardized, traceable, cost-effective and safe manufacturing of conditioned engineered tissue-like substitutes. This could lead to the successful translation of TE techniques as a daily clinical routine, defining an adequate correspondence between clinical efficacy and overall costs. Furthermore, quite recently, bioreactors have shown to be able to overcome limitations of conventional manual methods in seeding cells into porous scaffolds [Ikada, 2006]. In addition, they constitute technological instruments to perform controlled studies aimed at selectively understanding the effects of specific biological, chemical or physical cues on cell behaviour, maintaining or varying specific environmental parameters (e.g., temperature, pO_2, pH, pCO_2, shear stress) within defined ranges.

To date, bioreactor technology has improved the performances of 3D cultivated tissues, enabling the growth of high-density cell constructs, allowing the formation of 3D organized tissues in which cell viability is preserved and higher long-term survival rate of the cultured cells is ensured. Moreover, by dynamic cultivation within bioreactors, the mechanical signals that regulate the intra-cellular pathways responsible for tissue development, maturation and functionalization, can be piloted. Bioreactors have also introduced the opportunity to non-invasively and non-destructively monitor the construct development in real time, with enormous benefits in terms of costs, time, and quality control. However, although after the design of the first bioreactor TE has improved immensely, bioreactors still have major limitations and several issues must be further investigated in order to optimize dynamic cell culture supporting large-scale and long-term cell viability maintenance, which is mandatory for efficient engineered tissue maturation [Martin and Vermette, 2005].

2.1. Bioreactor Design Requirements

The design of a bioreactor is a complex task as the understanding of both engineering and scientific backgrounds are required in order to develop such an environment for the growth of engineered tissues. A number of criteria establish a blueprint for the design of a bioreactor. The design and the functional requirements of the tissue to be engineered determine the specific design requirements of the bioreactor, but in a general survey, a bioreactor must be designed to meet the following requirements [Lyons and Pandit, 2005]:

(i) to control the physiochemical environment
(ii) to facilitate monitoring of cell/tissue quality
(iii) to ensure the culture of tissue samples occurs under sterile conditions
(iv) to establish a substantial level of cellular distribution and attachment to developing scaffolds
(v) to ensure tissues sufficient nutrient and waste exchange with their surroundings (i.e. to provide efficient mass transfer to the tissue)
(vi) to expose the developing tissue to mechanical forces such as compression and expansion, as well as hydrodynamic forces such as shear stress and pressure
(vii) to maintain a high degree of reproducibility to control the flow of the media (steady/pulsatile)
(viii) to reduce excessive turbulence in the fluid flow
(ix) to provide a low volume capacity
(x) to make effective use of growth factors and medium components
(xi) to ensure that the materials from which the bioreactor is fabricated are compatible with cells/tissues
(xii) to be easy to clean and maintain
(xiii) to enable the user to easily fix the seeded scaffold in place
(xiv) to ensure the culture of tissue samples under physiological conditions
(xv) to be compact in size to fit in a standard size incubator
(xvi) to avoid the accumulation of metabolites

Hence, in designing bioreactors, both biomechanical and biochemical control are essential in creating a simulated physiological environment. Table 1 summarizes the ideal parameters to be controlled, i.e., required in the design of a bioreactor, in an attempt to mimic the parameters that exist *in vivo*.

Table 1. Ideal biochemical and biomechanical parameters to be controlled during dynamic cultivation in bioreactors

Biochemical Control	Biomechanical Control
Nutrient Delivery	Flow Rate
Po_2 And Pco_2	Volume
Waste Products Removal	Shear Stress
Ph	Pressure
Temperature	Resistance
Humidity	Compliance

2.2. Perfusion Bioreactors: Mass Transport Issue

The cultivation of 3D tissue constructs places great demands on the mass transport for the cell nutrient delivery. Tissues in the body overcome issues of oxygen and nutrient supply by containing spaced capillaries that provide conduits for convective transport of nutrients and waste products to and from the tissues. The 3D cell constructs that are developed *ex-vivo* or *in vitro* usually lack the vascular network that exists in normal vascularized tissues. As a result, in static cultures, the gas (O_2 and CO_2) and nutrient supply to the scaffold-seeded cells depends merely on mass diffusion. In static cultivation, with no fluid mixing, large diffusional gradients are formed between the cell constructs and their surroundings so that the cells in centre of the construct do not get sufficient nutrients, resulting in anaerobic cell metabolism; moreover, the waste removal from the centre is poor and thus cells encompass necrosis. This traditional approach does not fulfil all the requirements for regeneration of functional tissues or organs, in which a good perfusion of high-density 3D cell culture is required. Oxygen transport is typically considered as the main limiting factor for the establishment of long-term 3D cell culture [Colton, 1995]. Lack of perfusion, which would guarantee nutrient support, greatly limits the thickness of producible-engineered tissues.

To improve mass transport, researchers have developed various design solutions. Perfusion, also named mass transport-based, bioreactors have been introduced, which portray different patterns of fluid dynamics within the breeding chamber. Such bioreactors possess the ability to supply media to penetrate the most profound sections of the developing tissue to prevent hypoxia and necrosis. In contrast to static culture, in mass transport-based bioreactors active continuous transport of nutrients and gases to the cells, by the act of medium perfusion, promotes a more efficient aerobic metabolism [Kofidis et al., 2003]. As for the oxygen delivery to cells, when exposed to hypoxic conditions, cells undergo apoptosis. *In vivo*, apoptopic cells would be digested by phagocytes; however, *in vitro* these cells may leak proteases, DNA and other cellular components, which may have a detrimental effect on the surrounding tissue [Lyons and Pandit, 2005]. Conversely, if oxygen is being supplied at a rate that is too high, its concentration can increase up to inhibitory levels [Carrier et al., 2002]. Therefore, it is important that the oxygen requirement of the tissue is met to preserve the integrity and reliability of the tissue/organ culture system. A basic fluid-dynamic cultivation vessel is the spinner flask, which is an agitated flask usually working at 50 revolutions per minute (rpm). In these vessels, the cell constructs are subjected to turbulently mixed fluid that provides a well-mixed environment around the cell constructs and minimizes the stagnant layer at their surface. Spinner flask may not, however, be the optimal cultivation vessel for cells, since the turbulent fluid flow at the surface of the engineered constructs is usually characterized by eddies that can be destructive for the seeded cells or ultimately provoke cell detachment from the scaffold [Shachar and Cohen, 2003]. A milder dynamic cultivation environment is achieved in the rotating cell culture systems (RCCS), which were developed by NASA program to study tissue and cellular engineering in a low-shear, non-turbulent, simulated microgravity environment [Spaulding et al., 1993]. In RCCS, the operating principles are: i) solid body rotation along the horizontal axis, which is characterized by extremely low cellular fluid-induced shear stress and ii) convective motion of the medium, homogenizing oxygen distribution within the breeding chamber. In these devices cellular oxygen delivery is increased if compared with static cultivation, characterized by pure diffusion of the dissolved gas. The resultant flow pattern within the RCCS is laminar, with almost non-destructive fluid-induced shear stress. Furthermore, in RCCS, the seeded cells are in a state of a continuous free fall, i.e., rendering simulated microgravity. Fluid mixing is generated by the settling of the cellularized constructs associated with oscillations, tumbling, wake formation, and vortex shedding. Each construct thus mixes the medium in its immediate environment, and convective mixing in the chamber results as a whole from wake-wake collisions. Carrier et al. showed that pO_2, pH and pCO_2 levels were maintained in the RCCS to a better extent and allowed an aerobic respiration for a large number of cells compared to the static vessel. Cultivation of cardiac cell constructs in RCCS produced engineered cardiac tissues with improved cellular density, cell metabolism and expression of muscle specific markers. In comparison, cultivation in the mixed flasks resulted in tissues with low expression of specific cardiac markers and higher hypertrophy index [Carrier et al., 2002]. Finally, avoiding the onset of fluid turbulence, which is responsible for cell damage, the microgravity condition in RCCS seems to be the condition of choice for 3D high-density cell cultures, with respect to bioreactors based on stirring or agitation mechanism, maintaining cells within a 3D, low-shear-stress, high-mass-transfer, and high-oxygenation environment.

2.3. Bioreactors to Provide Mechanical Stimuli

It has long been recognized that external mechanical signals regulate the *in vivo* development and functions of tissues, according to a process named mechano-transduction. Cells in a 3D network respond to external signals via receptor-ligand interactions that relay such signals from outside the cell to the cytoskeleton domain and thereby influence subsequent cellular functions as attachment, migration, differentiation and apoptosis, ECM protein synthesis or gene expression. A large number of investigators worldwide have been working to delineate the molecular mechanisms accountable for the response of individual cells to mechanical stimuli [Shachar and Cohen, 2003]. Some tissues naturally exist in a mechanically dynamic environment. In terms of development and maintenance of the tissues and organs, our body is the "bioreactor", exposing the cell and ECM microenvironments to biomechanical forces and biochemical signals. In soft musculoskeletal tissue, external mechanical forces lead to adaptive remodelling (strengthening in response to exercise) [Ikada, 2006]. Cartilage, bone, ligaments, cardiac muscle and blood vessels resist against mechanical loading and hydrostatic pressure variation. For such tissues, it is therefore reasonably to assume that mechanical stimuli (pulsed or un-pulsed) should be applied to the cell-scaffold constructs *in vitro* with the aim to provide a more physiologic-like environment for tissue functionalization. As an evidence of this hypothesis, Zimmermann et al. and Eschenhagen et al. subjected circular molds of neonatal rat cardiomyocytes mixed with Matrigel, to a phasic mechanical stretch and this external imposed loading induced their contractility and the expression of specific cardiomyocyte markers [Zimmermann et al., 2002; Eschenhagen et al., 2002]. A mechanical stretch regimen has been also successfully applied in engineering cartilage, bone and skeletal muscle. Piccinini et al. exposed engineered cartilage tissues at different developmental stages to controlled loading regimes resembling a mild post-operative rehabilitation. Results indicated that the response was positively correlated with the amount of glycosaminoglycans in the constructs, suggesting that the engineered tissue could be better suited for early post-operative loading after implantation [Piccinini et al., 2009]. Another work has shown that pulsatile conditioning leads to improved properties of the engineered blood vessel, in terms of mechanical strength and cellular organization [Bilodeau and Mantovani, 2006]. These examples prove that external guides and signals, such as mechanical stress and strain, are essential to make cells *in vitro* grow into a functional 3D engineered tissue. However, these guides are difficult to apply directly to cells without any structural support. Therefore, the last strategy consists of using a scaffold, housed within the bioreactor, because this offers the possibility of an easier application of mechanical constraints on the young and fragile construct at the beginning of the regeneration.

Furthermore, given that mechanical stresses and strains are essential for the differentiation and the mechanical properties of every kind of tissue, the magnitude and the type of these stresses and strains are specific to cultivated tissues: vital mechanical stresses that are beneficial for cartilage culture would damage cardiac tissue [Bilodeau and Mantovani, 2006]. If the induced signals are inappropriate or absent, cells cannot proliferate and form organized tissues; they dedifferentiate and become disorganized, which can eventually lead to cell death or total inefficacy of the engineered tissue constructs. There are two major methods to mechanically stimulate cells in their culturing environment. Bioprocessing, which is the application of mechanical stimuli at intervals during the culture or a constant mechanical force to stimulate the cells, can be applied by i) stretching the substrate on which cells adhere, ii) by direct compression or iii) by imposing hydrostatic pressure gradients [Ikada, 2006]. The first and second stimuli are particularly suited for musculoskeletal and cardiac cells [Zimmermann et al., 2002; Eschenhagen et al., 2002] or osteoblasts [Ignatius et al., 2005] while the latter accounts for stimulating cartilaginous tissue [Elder and Athanasiou, 2009] or engineered heart valves [Hoerstrup et al., 2000]. In the case of engineering blood vessels, mechanical stimuli can be provided via pulsatile radial distension of circular tubes acting as scaffolds. The exposure of cells to microgravity can be also considered as a sort of mechanical stimuli, since fluid-induced shear stress acts at the fluid-cell interface on the cellular body.

In conclusion, several bioreactor-based TE applications have been proposed, but many more are expected. In table 2, bioreactors used for the cellular conditioning of different *in vitro* engineered tissues are revised Also, large-scale bioreactors for mass production of regenerated tissues and organs have already been proposed by a few industries in the field (Synthecon Inc.; BOSE Inc.; TGT Inc.; Aastrom Biosciences Inc.; Baxter International; FiberCell Systems, Inc.; Cesco Bioengineering Co., Ltd.; GE Heatlhcare; FlexCell Inc.).

Table 2. Bioreactors for *in vitro* cellular conditioning of engineered tissues. PGA: poly-glicolic acid; PLGA: poly-(lactic-co-glycolic acid); PCL: polycaprolactone

Authors	Engineered Tissue	Cell Source	Scaffold	Type of Bioreactor	Flow	Mechanical stimulus
Freed et al., 1997	Cartilage	femoropatellar grooves of bovine calves	PGA	rotating-wall	medium perfusion	negligible
Hoerstrup et al., 2000	Heart valve	endothelial cells and myofibroblasts from lamb carotid arteries	PGA	fixed-wall	pulsatile perfusion	cyclic mechanical stimulation induced by the fluid flow through the valve
Fink et al., 2000	Myocardium	cardiac myocytes of embryonic chicks and neonatal rats	collagen gel	fixed-wall	-	uniaxial longitudinal stretch
Gooch et al., 2001	Knee joint	bovine calves	PGA	rotating-wall	fluid mixing	shear stress induced by fluid mixing
Zimmermann et al., 2002	Myocardium	cardiac myocytes from neonatal rats	collagen gel	fixed-wall	-	phasic mechanical stretch
Altman et al., 2002	Ligament	human bone marrow stromal cells	silk fiber matrices	fixed-wall	pulsatile medium perfusion	translational and rotational strain
Bursac et al., 2003	Myocardium	neonatal rat cardiomyocytes	laminin-coated PGA mesh	rotating-wall	medium perfusion	negligible
Seidel et al., 2004	Cartilage	bovine calf articular chondrocytes	non woven PGA mesh	fixed-wall	medium perfusion	free-swelling, static and dynamic compression
Yu et al., 2004	Bone	rat osteoblastic cells	PLGA	rotating-wall	induced by rotation	shear stress induced by free-falling condition
Ignatius et al., 2005	Bone	human fetal osteoblastic cells	collagen gel	fixed-wall	-	pulsatile uniaxial stress
Braccini et al. 2005	Bone	bone marrow stromal cells	porous hydroxyapatite ceramic scaffold	perfusion bioreactor	medium perfusion	negligible
Kurpinski and Li, 2007	-	human mesenchymal stem cells	elastic silicone membrane	fixed wall	-	uniaxial cyclic tensile strain
Laganà et al., 2008	Cartilage	bovine chondrocytes from femoropoatellar grooves	porous scaffold	perfusion bioreactor	medium perfusion	shear stress combined with hydrostatic pressure
Jagodzinski et al., 2008	Bone	human bone marrow stromal cells	bovine spongiosa disc	perfusion bioreactor	medium perfusion	continuous perfusion and cyclic compression
Schulz et al., 2008	Cartilage	porcine chondrocytes	agarose hydrogel	fixed-wall	-	uniaxial stretch

Table 2. (Continued)

Authors	Engineered Tissue	Cell Source	Scaffold	Type of Bioreactor	Flow	Mechanical stimulus
Partap et al. 2009	Bone	osteoblastic cells	collagen-glycosaminoglycan scaffolds	perfusion bioreactor	medium perfusion	shear stress induced by fluid flow
Breen et al., 2009	Blood vessel	human umbilical vein endothelial cells	-	fixed wall	-	steady wall shear stress combined with cyclic tensile strain
Du et al., 2009	Bone	mouse osteoblasts	ceramic scaffold	perfusion bioreactor	medium perfusion	unidirectional / oscillatory flow
Ritchie et al., 2009	Esophagus	esophageal smooth muscle cells	polyurethane membranes	fixed wall	-	cyclic uniaxial strain
Gemmiti and Guldberg, 2009	Cartilage	cartilaginous tissue constructs	-	fixed wall	-	shear stress
Doroski et al., 2010	Tendon/Ligament	human marrow stromal cells	hydrogel cell carriers	fixed wall	.	cyclic strain
Barzilla et al., 2010	Mitral valve leaflet segment	-	-	rotating bioreactor	pulsatile medium perfusion	cyclic mechanical stimulation induced by the fluid flow through the valve
Ballyns et al., 2010	Meniscus	Bovine meniscal fibrochondrocytes	alginate carriers	spinner flask	medium mixing	tension/compression
Lee et al., 2010	Achilles' tendon	Tenocytes from rabbit Achilles tendon	PCL scaffold	fixed wall	-	tensile stretching
Kelm et al., 2010	Blood vessel	human artery-derived fibroblasts and endothelial cells	-	fixed wall	pulsatile flow	circumferential mechanical stimulation
Demenoudis and Missirlis, 2010	Blood vessel	endothelial cells	-	rotating wall	-	normal, circumferential and shear stress
Durst et al., 2010	Aortic valve	-	-	fixed wall	pulsatile flow	cyclic mechanical stimulation induced by the flow through the valve
Vismara et al., 2010	Blood vessel	Decellularized valve homografts	-	fixed wall	pulsatile flow	strain induced by pulsatile pressure load

2.4. Bioreactors as Organ Assist Devices

A further purpose in RM is the design and building of systems suited to directly support patients suffering of acute organ failure. In such applications, temporary treatment with bioreactors used as organ assist device, also named bioartificial hybrid bioreactors, allows the native organ to regenerate its functions, avoiding the need for transplantation and the resulting life-long immunosuppressive therapy.

In an effort to provide both longevity and improved quality of life for patients with chronic or end-stage organ failure, bioreactor support devices can be used also as a bridge to transplantation (replacing or partially supporting the functions of the diseased organ while waiting for an organ to be available for transplant), as well as in patients with early post-transplant dysfunction (until the complete restoration of the organ functionality) [Tzanakakis et al., 2000; Saito et al., 2006]. Regenerative temporary support can represent a particularly helpful strategy for the treatment of those acute diseases of organs which are relatively complicated to be engineered both *in vitro* or *ex vivo*. This approach is suitable for those organs made by several types of cells organized into functional microenvironments composed by several units having different roles, as for liver and kidney [Saito et al., 2006]. Indeed, although middle or late kidney precursor cells may develop into a partial organ or a tissue *in vitro*, these cannot be used for the repair of a diseased kidney [Dekel et al., 2003]. In the last few years, coupling the use of viable cells with the traditional hemofiltration technique used in dialysis treatment has lead to the development of bioreactors for extracorporeal kidney support. Within those devices, epithelial renal cells are cultivated adherent to hollow-fiber membranes housed within a synthetic tubular device. The blood filtrate obtained from the patient is conducted to the inside of the hollow-fibers and so exposed to the epithelial cell action. Hence, the blood is subjected to the natural filtration functions performed by the cells before being returned to the patient. To date, the clinical application of bioreactors for extracorporeal kidney support represents a better renal function replacement therapy than traditional dialysis therapy [Saito et al., 2006]. The enormous advantage of such devices is the possibility of integrating traditional hemofiltration with the full metabolic function of the renal cells in purifying and filtrate the blood. However, the treatment still suffers from the limitation of needing hospitalization of the patient. The next generation of kidney support devices is expected to be more compact and also wearable, for a continuous homemade hemofiltration therapy. Also the treatment of acute liver failure has evolved to the current concept of hybrid bioartificial liver (BAL) support, since engineered wholly artificial or bioartificial systems have not proven efficacious. BALs have been designed to support or replace the detoxification functions performed by the native liver. A first evidence of efficacy of this treatment is that BAL support allows minimizing the central nervous system complications (hepatic encephalopathy) related to acute liver failure states. The development of a BAL device involves many design considerations. First, the cells used in the device must be given an anchorage substrate and must be isolated from the host immune system to maintain viability and differentiated function. In BAL devices based on hollow-fiber configuration, which currently show the most promising efficacy [Park et al., 2005], the membranes separating compartments within the device must allow for adequate mass transfer, must provide immunoisolation, and must be biocompatible. Considering mass transport and nutrient delivery, despite several design solutions have been proposed, perfusion RCCS appears to be the most efficient solution in addressing the previous requirements [Wurm et al., 2007]. In BALs for hepatic support, maximizing oxygen and nutrient delivery to the hepatocytes is critical for optimal functioning, in order to not induce severe performance decrease for such devices [Catapano, 1996; Park et al., 2005].

3. ENABLING TECHNOLOGIES IN TISSUE ENGINEERING

Although impressive progress has been achieved in regenerating diseased tissues, it appears that traditional techniques employed in the RM field are inadequate to reach the production of wholly functional bioartificial 3D human tissues and organs. The problem is that several key questions concerning biological systems and their interactions with bioartificial materials and devices have not yet been still adequately answered, representing the "dark matter" in the experimental approach. Major drawbacks must be overcome before routine therapeutic application of engineered tissues can be realized [Solanki et al., 2008]. These include, primarily, the development of advanced techniques to better understand, and subsequently control, signalling in the biological microenvironments during *in vitro* cell culture and tissue maturation. As discussed previously, continuous development in technology and advance in life science knowledge have suggested that the application of bioreactor-based dynamic cultivation in TE should be able to address aspects of these challenges. Nevertheless, this question still remains unanswered: what are the conditions under which cells can be optimally cultured, enhancing interactions to enable the formation of a physiological-like functional tissue? Answers are needed because an optimized culture strategy would allow regenerating the tissue/organ in a more efficient way. Historically, culture conditions for constructs cultured in bioreactors (medium perfusion flow rate, O_2 tension, growth factor concentration, strain, shear stress, etc.) were "arbitrarily" selected prior to experiments. Moreover, conditions were typically maintained constant during the whole cultivation period (commonly between four and eight weeks) and then the construct was removed and analyzed. This trial and error approach has been expensive, inefficient and time consuming.

An innovative paradigm was proposed recently: culture conditions should be adapted to the maturation state of the engineered construct during the cultivation process. *In vitro* generation of a functional tissue requires maintenance of a suitable physiological environment that promotes efficient cell growth and proliferation while enabling correctly physiological activities. A living system is less a hierarchical, sequential transformational system, but more a highly concurrent, reactive system, with its own dynamic. A dynamic *in vitro* culture methodology that allows the tissue engineers to vary environment conditions and to zoom back and forth between lower and higher level behaviours during experiments is an attractive approach to studying emergence and complex phenomena within biological systems. Formation of a functional tissue is a dynamic process during which cell necessities change significantly as a consequence of dynamic cell re-organization within the construct. Accordingly, the cell-culture environment and the conditioning procedures should be changed concurrently with the cellular demands. For example, as the cell density increases because of cell proliferation, the initial imposed nutrient and gas concentrations and/or medium perfusion rate should be consistently increased during the cultivation period. Furthermore, because of gradual cell growth and maturation, synthetic scaffold degradation, new ECM synthesis and deposition, which alter mechanical properties, tuning of the cellular mechanical conditioning should be considered with respect to the initial imposed stimulation pattern [Sengers et al.,

2005]. Also, in stem cell culture, varying the cell conditioning procedure at precise stages can encourage cellular differentiation toward the desired phenotype.

Bioreactors can be used to recreate this dynamic environment, but when it comes to cell culture, the complex interplay of molecular and cellular factors makes identification and control of operating principles difficult. Recently, *in silico* modelling has been introduced in TE, as a tool by means of which supplement, extend and/or assist the development of time- and cost-effective research strategies for the prediction of suitable cellular and tissue conditioning procedures. In particular, multiscale modelling methods are now seen as being a reliable tool enabling evaluation of i) the cellular response to the conditioning procedures within bioreactors, along with predicting the dynamic modification of the cellular behaviour during the culture stage as a function of the external stimuli, and of ii) the most relevant features for optimizing bioreactor working conditions (i.e., the external stimuli for the cells), in terms of interactions between cells and the bioartificial hosting environment.

Increased use of computational simulation methods in TE has lead to establishment of *in silico bioreactors*. Simulating TE experiments could be addressed by creating a software framework able to incorporate a variety of phenomena influencing tissue growth along with a model of cell population dynamics. Predictive control may be achievable using such methods. The approach reduces reliance on traditional wet-lab trial and error methods, which are expensive and time-consuming. The approach can provide reliable information about how tissue maturation events emerge. It permits one to discover parameterizations capable of determining desired cellular responses to the *in vitro* conditioning while reducing uncertainties and reliance on physical experiments. *In silico* bioreactors have great potential. We envision that, in time, they will replace several aspects of current *in vitro* methodologies, overcoming *in vitro* limitations in providing the concrete answers required for moving to the *in vivo* stage. *In silico* bioreactors can outsource experiments designed to answer important practical questions. We believe that specific target identification and target validation are key elements for the future success of TE methodologies.

3.1. Multiscale in silico Modelling in Tissue Engineering

Substantial progress has been made in developing increasingly complex tissue engineered structures. Further progress is expected as a result of continued development and integration of several classes of enabling tools that will allow predictive and quantitative analysis and to begin characterization of the whole parameter space affecting new tissue formation and functionalization *in vitro*. All are involved in matching proper design requirements for cell culture experiments, and that requires analysis and investigation of phenomena at multiple levels. A multiple-scale approach is biomimetic. Consequently, in TE, moving from the macro to the nanoscale at increasing levels of detail is required in order to establish an effective design approach, to plan activities and strategies for the development of bioreactors, as well as for modelling and testing bioartificial engineered or native tissues functioning. Moreover, so doing enables the analysis of the interactions between cells/tissues and the engineered environments.

Multiscale modelling is a systematic approach allowing for selective collection of data to be incorporated in the global design scheme. The data are extracted from a complex system by investigating each variable and its effect individually and independently from the whole system behaviour. In multiscale models it is possible to observe more closely the behaviour of the system and to make predictions regarding its performance under altered input conditions and different system parameters [Consolo et al., 2010]. Faced with the task of gaining insight into cellular and biological phenomena, it is indeed very useful to split the system into sub-components, extract from them the most essential features, and than use that information to create a simplified but realistic system representation. The result is a "model" of that system. By explicitly taking advantage of the separation of scales or layers, the methods become tractable and much more efficient than solving the full fine-scale problem [Weinan and Engquist, 2003]. To date, modelling and multiscale approach are widely used in biomechanics. They join physical models, which use real constructions, with mathematical/ numerical models, which use conceptual representations. Multiscale modelling can be applied in TE to assist the process of designing innovative solutions and technologies or in optimizing product yield while gaining insight into the requirements of the cell culture environment. However, because of the complexity of the phenomena, analytical solutions are infeasible, so the introduction of advanced computational modelling and simulation methods is required to begin discovering for plausible biomechanical "laws" that govern the new-tissue formation. Computational advances coupled with increased interdisciplinary collaboration and the introduction of computer-aided design (CAD) methods has evolved into what is now called *computer-aided-tissue-engineering* (CATE). Advanced medical imaging along with modern design and manufacturing procedures have assisted TE development and created new directions for further TE advance. For example, using non-invasive computed tomography (CT) or magnetic resonance imaging (MRI) techniques to generate tissue structural views in support of realistic 3D anatomical models, or use of computer-aided design/manufacturing (CAD/CAM) and rapid prototyping (RP) technology to fabricate physical models of hard tissues, tissue scaffolds, and custom-made tissue implant prostheses [Sun and Lal, 2002]. However, within this chapter, CATE methods are considered primarily for utilization and integration of numerical and computational tools in modelling, investigating, and predicting TE structure behaviour.

Quantitative prediction through multiphysic and multiscale *in silico* modelling has been demonstrated to have a critical role in the evolution of TE, enabling rational design and successful realization of engineered tissues [Pancrazio et al., 2007]. *In silico* models can supplement and extend the current empirical TE and RM techniques. They facilitate developing cost-effective strategies for predicting patterns and time lines of the new-tissue formation and/or tissue regeneration. They also place in the hands of researchers methods by which specific hypothesis can be evaluated, tested and validated or falsified. A practical enabling feature is that they expedite testing of new constructs, scaffolds and devices, and prototypes. Because of the ability of *in silico* TE to enable high-throughput experiments, we view it as a diverse, active specialty that can lead efforts in overcoming many of the current RM and TE limitations. Achieving the vision requires integration of multiphysic and multiscale computational technology into the traditional design and manufacturing process.

Within the next sections i) computational fluid dynamics (CFD) modelling and ii) multi agent systems (MAS) are presented. Potential of the two methods are highlighted in allowing prediction and extraction of reliable settings for a rationale TE experiment design.

3.2. CFD-Based Modelling Technique in Tissue Engineering

In the following sections computational fluid-dynamics (CFD) techniques, allowing for numerical modelling and simulation of phenomena in which fluid mechanics aspects are involved and controlled fluid dynamic conditions are essential, are described and evidence is provided relative to the efficacy of the CFD-based method in enabling the prediction of a large number of parameters that influence cell growth in a tissue engineering context. Computational methods can be a primary instrument in modelling and design TE experiments with controlled fluid-dynamic conditions. Indeed, characterization of the complete 3D fluid flow or the identification of optimal flow conditions within perfusion/mass transport bioreactors are too time consuming to be determined by experimental velocimetry techniques but rather should be predicted by simulation methods [Martin et al., 2004]. CFD modelling provides a powerful means for overcoming these limitations.

Essentially, CFD modelling is a technique based on the development of numerical methods and algorithms allowing solving and analyzing problems in which fluid mechanics aspects are involved. As a first result, flow characteristics within a porous scaffold or a perfusion dynamic bioreactor can be fully investigated at multiple scale level, also accounting for the evaluation of the cell-environment interactions. Scaffolds and bioreactor design can thus be theoretically evaluated and characterized before fabrication. Bioreactor specific operating conditions, such as fluid inlet velocity or the fluid-induced cellular shear stress, can also be selectively varied in the *in silico* model to better predict their influence on cell growth. Moreover, CFD codes generally enable visualization of flow phenomena, an important benefit when it is impractical to position probes within the fluid domains for the measurement of parameters such as pressure and velocity. Simulations by several groups demonstrated how the scaffold shape and morphology influence hydrodynamic shear experienced by cells and nutrient delivery: in an attempt to better characterize the local hydrodynamic environment seen by the cells (i.e., within the scaffold pores) multiscale computational models have been developed, based on (idealized or precise imaging-based) reconstruction of the scaffold geometry and pore architecture [Cioffi et al., 2006; Galbusera et al., 2007; Cioffi et al., 2008; Cantini et al., 2009]. Numerical techniques are effective in capturing the flow, pressure and concentration fields at the pore level of a scaffold. By CFD-based modelling technique, an optimized 3D geometry and micro-nanoarchitecture can be derived, for a reliable design of the scaffolds. Moreover, the interactions between scaffold pore morphology, mechanical stimuli developed at the cell microscopic level, and culture conditions applied at the macroscopic scale can be studied using a CFD-based computational approach, as recently proposed by Olivares et al. (2009). Also, in bioreactor design, the need for precision control together with the need of evaluating the effect of each specific parameter for tissue conditioning has lead to the development of several multiscale-based methods, supported by numerical and computational modelling.

As reported in Semple et al., the use of computers as *in silico* bioreactors, that (at least) partially replace laboratory technology as the environment in which engineering functional tissues, is one of the key areas where the future of TE resides [Semple et al., 2005]. The fundamental design of operating conditions of bioreactors should be founded on a rational design process that can be more easily achieved by means of computational modelling. The use of computational methods to predict and understand the flow-dependent processes in bioreactors can not only improve the overall performance of the device, but also will likely reduces both the time and costs of manufacture. Multiscale computational models have been developed in recent years, which simulated the fluid dynamics and mass transport at the level of the macro- and microscale environment in mass transport-based bioreactors [Williams et al, 2002; Sengers et al., 2005; Mareels et al., 2006; Raimondi et al., 2006; Kasper et al., 2009; Consolo et al., 2009]. Various features related to the flow field characteristics within bioreactors can be analyzed by means of CFD-based modelling, which are: i) gas (i.e., oxygen (O_2) and carbon dioxide (CO_2)) and nutrient delivery to the viable cells (transport phenomena), ii) cellular O_2 consumption and catabolite production rates (cellular metabolic activities), and iii) hydrodynamic forces (wall shear stress and fluid-viscous mechanical stress) experienced by cells within dynamic environments. The design of a device optimizing these aspects is essential, since the main challenge for *in vitro* successful high-density cell culture under perfusion is the trade-off between a high supply of nutrients and a not damaging induced cell stress. In such applications, characterizing and quantifying the flow velocity profiles, mechanical stresses, possibly together with the transport of specific biochemical species in the local environment of the cells is of primary importance. In particular, regarding O_2 supply, quantifying O_2 levels within perfused constructs is not only necessary to predict whether specific design solutions can supply sufficient O_2 levels to maintain cell viability and functions. Indeed, the differentiation stage of stem cells can be modulated by applying different ranges of O_2 during culture (e.g., hypoxia: 1%-5% O_2 vs normoxia: 20% O_2) [Gassmann et al., 1996; Malda et al., 2004]. Thus, it may be important to identify proper conditions that can supply cells with a defined range of O_2, as a potential strategy to modulate cell differentiation toward the desired phenotype. Computational models also helped to tune dynamic bioreactor working conditions in order to avoid the onset of fluid turbulences causing fluid-induced cell mechanical trauma, i.e., the generation of cell-damaging hydrodynamic forces during cell culture [Consolo et al., 2009]. Furthermore, in applications where cells are cultured adhered to carriers or encapsulated within polymeric beads, numerical simulations have been performed, in conjunction with experiments, to verify the trajectory of carrier/beads within rotating bioreactors [Pollack et al., 2000].

3.2.1. CFD: Governing Equations

The solution of a CFD model is fundamentally based on the numerical solution of the Navier-Stokes equations, which are differential equations describing the motion of a fluid phase [Ferziger and Peric, 1996]. These equations arise from applying Newton's second law to fluid motion, together with the assumption that the fluid stress is the sum of a diffusing viscous term (proportional to the gradient of velocity), plus a pressure term. The Navier–Stokes equations in their full and simplified form are reported, which are the mass continuity equation (eq. 1), a statement of the conservation of mass, and momentum conservation equation
(eq. 2):

Conservation of Mass: $$\frac{\partial \rho}{\partial t} + \nabla \cdot (\rho \vec{v}) = 0 \qquad \text{(eq. 1)}$$

Conservation of Momentum: $$\rho\left(\frac{\partial \vec{v}}{\partial t} + \vec{v} \cdot \nabla \vec{v}\right) = -\nabla P + T + F \qquad \text{(eq.2)}$$

where ρ, $\vec{v}$ and P are respectively the fluid density, the velocity vector and the pressure, T is the deviatoric stress tensor, and F represents the body forces (per unit of volume) acting on the fluid. In eq. 2, the left side of the equation describes acceleration, and may be composed of time dependent or convective effects, while the right side is in effect a summation of body forces (such as gravity) and divergence of stress (pressure and stress). The Navier–Stokes equations result from a eulerian approach in describing the fluid motion: Navier–Stokes equations dictate not position but rather velocity. A solution of the Navier–Stokes equations is called a velocity field or flow field, which is a description of the velocity of the fluid at a given point in space and time [Ferziger and Peric, 1996]. In many applications, other equations are solved simultaneously with the Navier-Stokes equations. For instance, in modelling nutrients supply to viable cells within mass transport-based bioreactors the one describing species concentration (mass transfer) counting chemical reactions can be included, which is reported in its vector form, in eq. 3:

Species Conservation of Mass:

$$\frac{\partial}{\partial t}(\rho Y_i)+\nabla\cdot(\rho \vec{v} Y_i)=-\vec{\nabla}\cdot \vec{J}_i+R_i+S_i \qquad \text{(eq. 3)}$$

When the conservation mass equation for chemical species (eq. 3) is included in CFD calculations, the local mass fraction Y_i is predicted, though the solution of a convective-diffusion equation for the ith species. In eq. 3, R_i represents the net rate of production of species *i* by chemical reaction and S_i is a mass source term from the dispersed phase. For laminar flow, the term $\vec{J}_i$ in eq. 3, which is the diffusion flux of species *i* arising due to gradient of concentration, can be expressed by the Fick's law, in the form reported in eq. 4:

Fick's Law: $$\vec{J}_i=-\rho D_{i,m}\nabla Y_i \qquad \text{(eq. 4)}$$

where $D_{i,m}$ is the mass diffusion coefficient for species *i* in the fluid bulk phase *m*. More advanced computational codes provide as well for the simulation of more complex cases involving multi-phase flows (e.g. liquid/gas, solid/gas, liquid/solid mixtures), non-Newtonian fluids (such as blood), or chemically reacting flows. For multi-phase flows, in which a primary phase *p* contains *n* dispersed granular secondary phase *s*, continuity and conservation of momentum equations (eq. 1 and 2) are written in the form reported in equations 5 and 6 (equations are described for the primary phase *p*, analogous equations can be written for the secondary phase *s*):

Conservation of Mass: $$\frac{\partial}{\partial t}(\alpha_p\rho_p)+\nabla\cdot(\alpha_p\rho_p \vec{v}_p)=\sum_{s=1}^{n}(\dot{m}_{sp}-\dot{m}_{ps})+S_p \qquad \text{(eq. 5)}$$

Conservation of Momentum:

$$\frac{\partial}{\partial t}(\alpha_p \rho_p \vec{v}_p) + \nabla \cdot (\alpha_p \rho_p \vec{v}_p \vec{v}_p) = -\alpha_p \nabla P + \nabla \cdot \bar{\tau}_p + \alpha_p \rho_p \vec{g} + \sum_{s=1}^{n} (\vec{R}_{sp} + \dot{m}_{sp} \vec{v}_{sp} - \dot{m}_{ps} \vec{v}_{ps}) + (\vec{F}_p + \vec{F}_{lift,p} + \vec{F}_{vm,p}) \quad \text{(eq. 6)}$$

where α_p, ρ_p and $\vec{v}_p$ are respectively the volumetric fraction, the density and the velocity vector of the primary phase p. The term $\dot{m}_{sp}$ characterizes the mass transfer from the sth to the pth phase while $\dot{m}_{ps}$ characterizes the mass transfer from phase p^{th} to the sth; S_p is a mass source term in the fluid bulk phase. $\bar{\tau}_p$ is the pth phase stress-strain tensor and $\vec{R}_{sp}$ is the interphase force, depending on the friction, pressure, cohesion, among the phases. $\vec{F}_p$ is an external body force, $\vec{F}_{lift,p}$ is a lift force, $\vec{F}_{vm,p}$ is a virtual mass force, and P is the pressure shared by all phases.

3.2.2. CFD Modelling: Spatial Discretization and Implementation of Numerical Simulations

The most fundamental consideration in CFD modelling is the discretization of a continuous fluid phase (also named as the fluid domain): one method is to discretize the spatial domain into small "cells" (uniform or non-uniform) to design a volume mesh or grid, and then apply a suitable algorithm to solve the equations of motion for the discretized fluid domain. For this purpose the Finite Volume Method (FVM) is the standard approach used most often in commercial software and research codes treating fluid phase problems [Versteeg and Malalasekra, 1995]. High-resolution discretization schemes allow for modelling a variety of geometries, even if spatial discontinuities are present in the fluid domain. In developing a CFD-based model, the following general procedure can be performed: the first step (i.e. the geometrical model design) includes:

- the definition of the geometry (physical bounds) of the problem
- the partition of the volume occupied by the fluid into discrete cells (the mesh)
- the definition of the physical problem (for example combining the equations of motions with those for enthalpy or species conservation calculations)
- the definition of the boundary and, for transient flows, of the initial conditions (this involves specifying the fluid behavior and properties at the boundaries of the problem and at the beginning of the numerical simulation

The next step is the running of the numerical simulation (equations, depending on the previous settings, are solved iteratively as a steady-state or transient) and finally a postprocessor is used for the analysis and visualization of the resulting solution. As for the solution algorithms, discretization in space produces a system of ordinary differential equations for unsteady problems and algebraic equations for steady problems. Implicit or semi-implicit methods are generally used to integrate the ordinary differential equations, producing a system of (usually) nonlinear algebraic equations [Versteeg and Malalasekra, 1995].

3.3. CFD-Based Solution Strategies for in silico Design of Scaffolds and Bioreactors

Computational modelling methodologies are extensively used in scaffold design, including numerical and computational methods for solving the fluid (and when necessary the solid) mechanics features involved in scaffold properties optimization. To date, several studies have also proved the usefulness of CFD-based modelling in analyzing fluid dynamics and transport of chemical species in bioreactor models. Interestingly, CFD modelling permits studies to be carried out at multiple-scale levels, which are: a) a macroscopic level, for the analysis of the comprehensive scaffold-cell interactions in correctly supporting and enhancing the engineered new-tissue development, b) a microscopic level, for the analysis of fluid dynamics inside the idealized or realistic 3D model of the scaffold architecture, and finally c) at the sub-microscale for the analysis of the cell-fluid interactions and of the cellular response to the cultivation/conditioning procedure. Multiscale CFD-based technique validly allows for modelling simultaneously all these issues [Consolo et al., 2010].

Regarding to scaffold design, by implementing a multiscale CFD-based model, significant clues for the design of the culture substrate can be retrieved. A CFD model allows porous scaffold morphology design to be optimized iteratively by single-parameter analysis, varying selectively in the model geometry, pore scaffold size, shape, and/or pore interconnections. By *in silico* modelling the fully developed flow inside the scaffold, the iterative optimization of the scaffold architecture may result from the evaluation of the influence of the scaffold geometry and morphology on the nutrient supply as well as on the distribution of shear stress on scaffold-adhered cell body.

As two case studies of the enormous potential of multiscale CFD-based modelling, applied to the TE field, we mention the study by Cantini et al., who used CFD modelling to optimize the microenvironment inside scaffolds for dynamic co-culture of hematopoietic stem cell (HSCs) and bone marrow stromal cells (BMSCs) in a perfusion bioreactor [Cantini et al., 2009], and the combined macro/micro-scale computational approach proposed by Cioffi et al., to quantify O_2 transport and flow-mediated shear stress to human chondrocytes cultured in 3D scaffolds in a perfusion bioreactor system [Cioffi et al., 2006]. The CFD geometrical model developed by Cantini et al. was aimed at the optimization of the scaffold geometry to avoid non-adherent stem cells being dragged away while ensuring adequate nutrient supply during perfusion in bioreactors, as a function of the fluid dynamic conditions imposed within the culture chamber. The study was based on the simplification of the geometry of a synthetic micro-porous scaffold manufactured using the solvent casting/particulate leaching technique (an experimental procedure of particulate leaching consists of manufacturing the template of a porous scaffold via packing and interconnecting solid particles (porogen), casting and evaporating a polymer solution over them, and finally removing the particles by leaching.). To analyze the flow through the void porous microstructure, the geometrical model developed consists of interconnected spheres representing the pores where culture medium flows and HSCs are homed, whereas BMSCs are assumed to be spread on the surface of the pores and therefore to coincide with the walls of the model. Fully developed flow inside the scaffold was modelled using periodic inlet and outlet boundaries, because of the periodically repetitive nature of the physical geometry of interest and of the expected pattern of the flow solution. Another hypothesis tested in this work is whether the addition of longitudinal micro-channels to a homogeneous porous scaffold is able to improve its in-depth perfusion by providing artificial capillaries for the supply of nutrients during culture. For the evaluation O_2 transport and O_2 cellular consumption, a different model was used, because a periodic description did not fit these phenomena straightforwardly. A non-periodic model was built, consisting of superimposed periodic units, to allow for the full height of a 3-mm-thick scaffold. The model accounted for calculation of shear stress on the walls of the reference case and of O_2 partial pressure (pO_2). In calculating pO_2, different geometries were considered and compared with those of a solely porous model (without channels) of the scaffold. The modification of the characteristics concerning the porous medium (diameter of the pores and porosity) mainly accounted for evaluating the influence of the scaffold design on the distribution of shear stresses on pore walls. The results of O_2 transport computations showed the effectiveness of the channels in improving O_2 delivery, which has been quantified in a 34% higher pO_2 at the outlet of the 3-mm-thick scaffold. In conclusion, the work by Cantini et al. provides the basis for a CFD-driven design of a dynamic culture system with micro-structured scaffold, acknowledging the increasingly important role of computational methods in defining optimal flow conditions for perfusion bioreactors.

The study by Cioffi et al., regards a multi-scale scaffold model based on precise reconstruction of the scaffold geometry by imaging methods, and, in particular, by micro-computer tomography (micro-CT) imaging method reconstruction, for the numerical simulation of the fluid-induced mechanical stimulation in scaffold for articular cartilage replacement where chondrocytes are cultured [Cioffi et al., 2006]. In chondrocyte-seeded constructs, new-ECM synthesis is particularly promoted by hydrodynamic flow, which results in cell membrane stretching, caused by fluid-induced shear, and in enhanced convective transport of nutrients to the cells and of catabolites away from the cells. The shear stress imposed to chondrocytes in such systems will depend not only on the culture medium flow rate through the constructs but also on the scaffold 3D architecture. A macroscale model has been first developed to assess the influence of the bioreactor design and to identify the proper boundary conditions for the microscale model. The microscale model, based on a micro-CT reconstruction of a synthetic-polymer foam scaffold, has been developed to assess the influence of the scaffold micro-architecture on local shear stress and O_2 levels within the scaffold pores. Furthermore, perfusion culture experiments were performed in the modelled cell–scaffold bioreactor system to derive specific O_2 consumption rates for the simulations. Once the model has been constructed, the flow field within the porous scaffold, the shear stress conditions experienced by seeded chondrocytes and the O_2 tension have been numerically evaluated at different flow-perfusion rates, by means of a computational software-based on the finite-volume technique. The study by Cioffi et al. confirms that computational modelling can be successfully used in combination with micro-imaging techniques, to quantify, on different scale levels, the shear stress induced and the delivery of O_2 to cells in 3D engineered cellular bioreactors: these predicted data provide insight into the potential mechanical interaction between the cells and their surrounding environment, and may be of practical use to investigate mechanisms of mechanical signal transduction. Furthermore, computational models may provide new scaffold design criteria, such as indications on the microstructure that could optimize transport phenomena to and from the cells.

In summary, the reported studies prove that multiscale CFD-based models may result in a valuable tool to *in silico* appreciate how a controlled variation of the geometry affects the overall scaffold design and functioning. Moreover, they provide clues to define the value of bioreactor working conditions, e.g., perfusion flow rate or rotation velocity in reactors working in simulated microgravity conditions [Horner et al., 1998; Sucosky, 2005; Singh et al., 2005; Sengers et al., 2005; Mareels et al., 2006; Boschetti et al., 2006; Galbusera et al., 2007; Ma et al., 2007].

We here report two studies by our group aimed at the optimization of transport phenomena and cell-environment interactions within two mass-transport bioreactors operating in simulated microgravity conditions [Consolo et al., 2009; Consolo et al., 2010b].

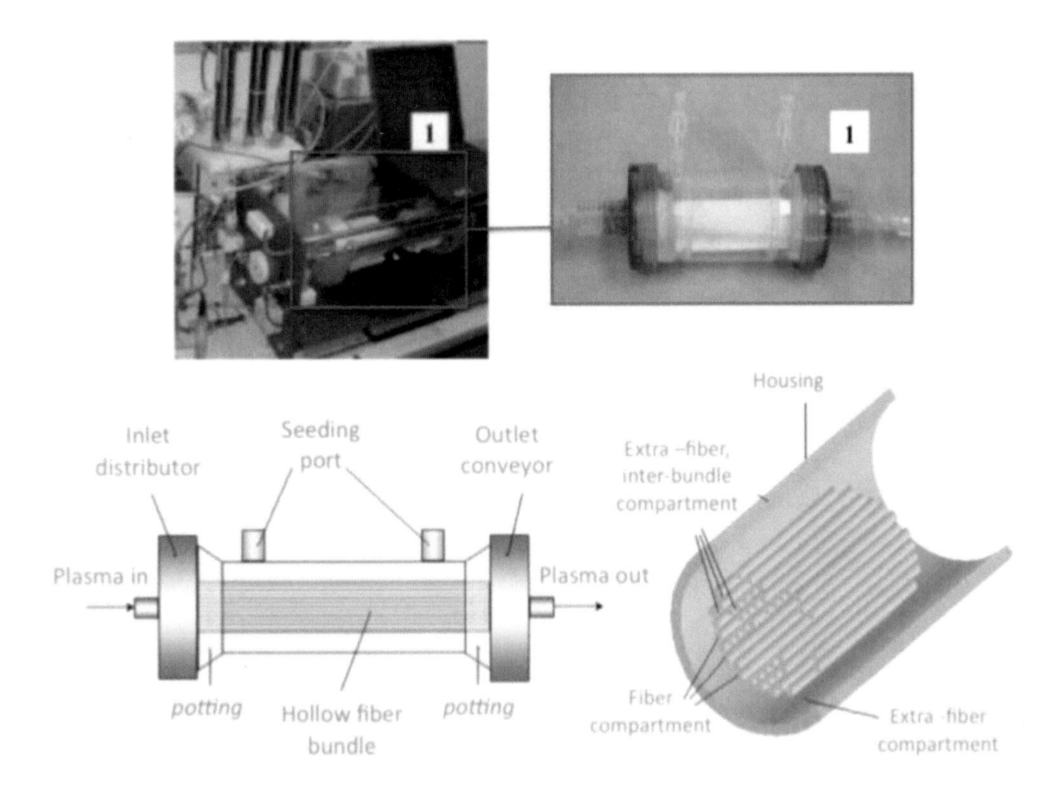

Figure 4. The prototype (top panel) and a schematic representation of the breeding chamber (bottom panel) of the bioartificial liver (BAL). The patient plasma is perfused, in an extra-corporeal circuit, through the hollow fibers while hepatocytes are cultivated in the extra-capillary space; the direct plasma–cell contact is due to plasma filtration in the cellular compartment, enhancing efficient mass transfer. The cutoff (pore diameters) of selective membranes of the fibers avoids the filtration of proteins such as immunoglobulins, preventing immune-mediated injuries due to the use of allogenic and xenogenic hepatocytes. Permeable membranes separate the hepatocytes from the flowing plasma, effectively protecting the cells from flow-induced shear stresses. Culturing microcarrier-attached cells greatly facilitate nutrients and waste transfer.

In the first study, a comprehensive computational study modelling the operation of a rotating hollow-fiber bioreactor for artificial liver (BAL) was performed to explore the interactions between the oxygenated culture medium and the cultured hepatocytes. The investigated BAL (fig. 4) combines the advantages of hollow-fiber-based devices with the advantages guaranteed by operating in microgravity. The investigated device is a hybrid BAL, owned by Fresenius Medical Care GmbH (Bad Homburg, Germany). The device, designed for culturing viable hepatocytes under simulated microgravity conditions, provides mass exchange by internal filtration through a bundle of hollow, selective, semi-permeable, porous fibers placed in the middle of the breeding chamber. This BAL is meant for culturing and growing suspension of anchored cells aimed at assisting the recovery of native organ functionality (bridge to recovery) or temporarily replacing it (bridge to transplantation) while supporting patients affected by end-stage hepatic diseases and is embedded in an extra-corporeal circulation circuit. Under *in vitro* conditions, culture medium flows through hollow fibers (the fiber compartment).

The medium is oxygenated using an external oxygenator before entering the fiber bundle, thus working as an O_2 fluid vector for the cultured hepatocytes. Adhering on the surface of microspherical carriers of Cytodex[3], cells float in the culture medium solution (the extra-fiber or cellular compartment), forming spheroidal aggregates with a mean diameter of 500μm. Such aggregates are composed of Cytodex[3] microcarriers connected to one another by means of microvillous protrusions on the surface of the clustered, attached hepatocytes (which fill the gaps between the microcarriers, fig. 5). In this study, we proposed a comprehensive numerical investigation providing a detailed analysis of the transport phenomena occurring inside the breeding chamber. A 2D model was first developed to investigate the effects of microgravity on suspension of micro-carrier-attached-aggregated-cells (μCAACs), in particular to search for the rotational speed value of the chamber that would ensure the appropriate homogenous distribution of aggregates within the fluid domain (fig. 5).

Figure 5. Left panel) contours of volume fraction (VF, expressed in percentage) of the μCAACs in the 2D model rotating at 5 rpm (a), 10 rpm (b), 20 rpm (c), 25 rpm (d), 30 rpm (e), and 35 rpm (f), after a 300-s simulated time. A rotating speed of 30 rpm allows homogeneous distribution of the floating μCAACs to be achieved. At lower speeds, sedimentation of aggregates is observed; at 35 rpm centrifugal effects influence cellular distribution. Right panel) spherical μCAACs (mean size equal to 500μm in diameter), which are formed by Cytodex[3] microcarriers (green) connected to one another by means of microvillous protrusions on the surface of the clustered, attached hepatocytes (red).

From our results, it has been possible to evaluate individually the effect of the variation of the BAL rotational speed on cellular aggregate motion and distribution, identifying that value (30 rpm) that guarantees the proper microgravity mixing of floating μCAACs. Then, a 3D model was developed to describe the mass-transfer interactions between the oxygenated culture medium and the cells (fig. 6); the implemented 3D model provides a global analysis of the functioning of the device, i) describing simultaneously the microgravity motion of the viable μCAACs, ii) analyzing the internal filtration of the culture medium through the hollow fibers, iii) predicting the delivery of O_2 to the cells and iv) the cellular O_2 consumption, and v) providing for calculation of the fluid-induced stress experienced by floating aggregates, integrating diffusion, convection, and multiphase fluid dynamics calculations. According to results from previous 2D analysis, simulations with the 3D model confirmed that a rotational speed of 30 rpm allows homogeneous distribution of the simulated μCAACs to be achieved, avoiding μCAACs settling onto the bottom device wall and the formation of excessively packed cellular clusters (fig. 6).

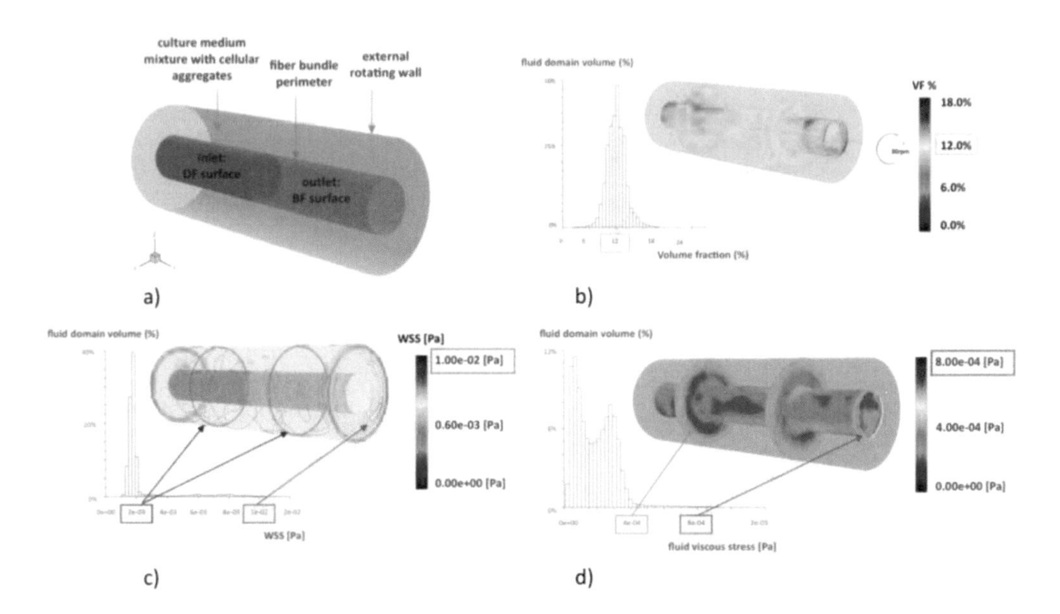

Figure 6. a) 3D CFD model of the fluid volume of the bioreactor. The annular region represents the fluid volume of the bioartificial liver where hepatocytes are suspended. Mass exchange between the fibers and the cellular compartment was modelled to occur at the interface surface of the central hollow cylinder, modelling the fiber bundle. b) contours and percentage volume distribution histogram of the volume fraction (VF, expressed in percentage) of the μCAACs within the 3D model plotted along interesting internal surfaces; distribution of the wall shear stress (c) and of the fluid-induced viscous stress (d) within the breeding chamber of the 3D model as experienced by floating cellular aggregates. Rotational speed = 30 rpm; 1200-s simulated time.

Moreover, a rotational speed of 30 rpm, in addition to maintaining μCAACs floating close to a gravity-free condition, can be considered optimal for culturing μCAACs within the rotating BAL, because it produces a non-damaging fluid-induced mechanical stress for cell specific-liver functions and enhances aggregation of microcarrier-anchored hepatocytes. Such optimized conditions permit microcarrier-attached cells to be uniformly suspended in the fluid with minimum mechanical stress. Concerning O_2 delivery, data provided by our analysis demonstrated that, setting the appropriate inlet flow rate and pO_2 values within the filtering medium, a condition in which pO_2 is higher than the critical threshold value (10 mmHg), causing a decrease in cell metabolic activity and hypoxic damage to cultured hepatocytes, can be attained within the BAL chamber (fig. 7).

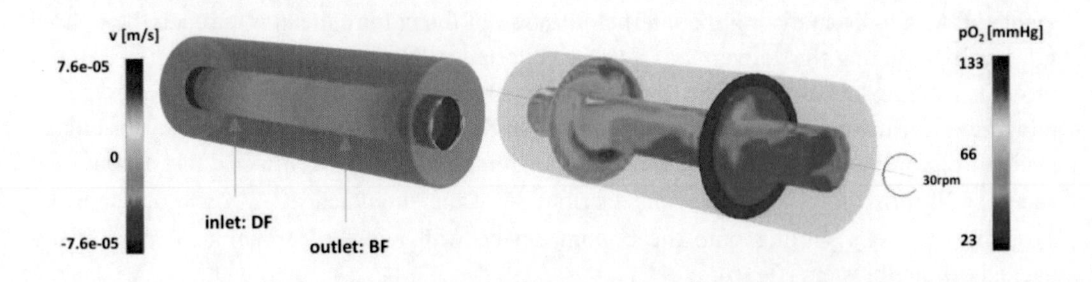

Figure 7. Contour maps of a) the radial medium ultra filtration profile assigned in the model (DF: direct filtration; BF: back filtration) and of b) pO_2 within the fluid domain of the 3D model after 1200-s of evolution time of the computational simulation.

In computing O_2 distribution within the chamber, liver-specific cellular O_2 consumption rates have been taken into account. In conclusion, we have demonstrated the feasibility of using computational tools to search for optimal conditions for culturing viable microcarrier-attached aggregated hepatocytes in a hollow-fiber/microgravity-based BAL. In the near future, this might become a key strategy to help overcome the challenging issues in optimizing O_2 supply, uniform cellular distribution, and maintenance of viability of aggregated hepatocytes in rotating BALs. In addition, the proposed approach could be extended to similar applications in which the determination of the requirements of bioartificial devices is strongly based on the fluid dynamic and mass transfer patterns.

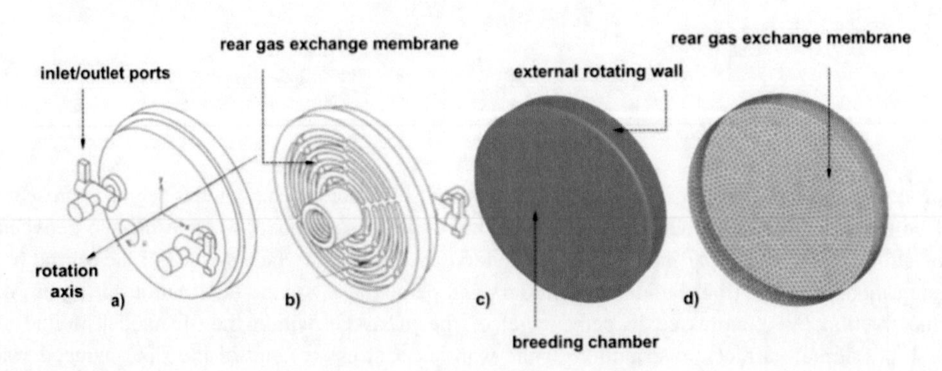

Figure 8. Schematic of the 55-mL disposable vessel (a, b), 3D computational fluid dynamic model (c), and detail of the 3D mesh of the gas exchange membrane (d) of the rotating breeding chamber of the HARV bioreactor in which alginate-bead encapsulated mESCs are suspended.

In a second study, a CFD-based modelling study for the computer-aided design of a scalable, automated and functional large-scale 3D bioprocess for the *in vitro* generation of differentiated embryonic stem cell-derived cardiomyocytes (ESC-DCs), within a 3D rotatory culture environment, was developed. The main of the *in silico* model was to simulate the *in vitro* cardiogenesis bioprocess, which was carried out suspending alginate-bead encapsulated murine ESCs (mESCs), in a pro-cardiogenic conditioning medium, inside the rotating breeding chamber of the High Aspect Ratio Vessel (HARV, Synthecon Inc,, USA; fig. 8).

In detail, the computational model was developed in order to extract the set of parameters optimizing suspension conditions of encapsulated cells within the HARV.

The computational approach allowed us to firstly develop a 2D model aimed at investigating the effect of the bioreactor rotation on suspension of alginate-encapsulated cells, in particular searching for the rotational velocity value of the chamber i) ensuring an appropriate homogenous distribution of alginate beads floating in simulated microgravity condition, and ii) avoiding that the alginate-carriers (i.e., the beads) do frequently collide with the boundary walls of the chamber, thus eliciting cell damage or bead rupture. Then, a fully-3D model was developed, also accounting for the evaluation of the mass-transfer interactions between the oxygenated culture medium and the cultured cells. In detail, the 3D computational model, based on multiphysic modelling, simultaneously describes i) the microgravity motion of the floating beads, ii) the delivery of O_2 to the cells, also iii) taking into account the cellular O_2 consumption, as a function of the physical rotation of the breeding chamber. According to results from numerical simulations, at a rotational speed of 25 rpm beads are quite-homogeneously distributed within the HARV (fig. 9): neither excessively packed beads nor deposited beads are identifiable and no relevant sedimentation is observed. In this established condition, the hydrodynamic force induced by the physical rotation of the chamber overcomes the effect of the gravitational force, thus ensuring a proper mixing of the suspended beads, which are maintained in a quasi-periodic circular motion. Moreover, encapsulated cells are mostly located away from the device external wall, so that potentially damaging bead-wall collisions are minimized. As for cell O_2 supply, notably, the imposed suspension condition attained when a rotational speed equal to 25 rpm is set maintains the floating encapsulated-cells close to a normoxia condition (pO_2 is in the range 157-159 mmHg; fig. 9) in the whole fluid domain so that no regions suffering from insufficient oxygenation are observed. Such a condition revealed to be favourable for high-density mESCs long-term viability maintenance, which is mandatory to promote mESCs differentiation toward the cardiac phenotype. Furthermore, this result is of primary importance, proving consistency of the identified device suspension conditions in overcoming unsolved issues in O_2 supply as well as in the generation of cell-damaging hydrodynamic forces, faced in conducing cell culture in static petri dish or within dynamic bioreactors based on the agitation/stirring mechanism (as for Spinner Flasks or Stirred Vessels), respectively. Results extracted from the computational analysis served as criteria to set the operating conditions for preliminary *in vitro* tests, conducted by culturing the encapsulated mESCs in the HARV dynamic environment for 21days. The set of suspension conditions derived from numerical predictions have been successfully tested for the in vitro generation of mESC-DCs (fig. 10).

Preliminary *in vitro* tests showed that at day 21 of cardiogenic culture extensive portions of cell aggregates sized 200μm, as an average, in diameter resulted highly viable, showing no necrotic core and a limited amount of dead cells of small dimension (fig. 10). Conversely, after 21 days of static culture carried in petri dish cell viability resulted dramatically decreased within alginate beads: as shown in figure 10, cell aggregates resulted of smaller size with respect to the dynamic condition, showing several necrotic cores whilst only the small colonies (≤ 40μm) resulted viable. As for cell differentiation, our results clearly show a gradual downregulation of undifferentiated markers whilst upregulating mesodermal genes, with final maturation of the cardiac lineage. The herein presented results demonstrate the effectiveness of i) the encapsulation-based aggregation control, ii) the rotatory bioreactor-based physiochemical conditioning, and iii) the use of the pro-cardiogenic conditioning medium in allowing the growth of size-controlled cellular clusters under simulated near free-gravity conditions, promoting the generation of a mesodermal pool of cells during mESC differentiation towards functional cardiomyocytes. Moreover, the feasibility of using CFD-based computational tools as reliable and cost-effective strategy to assist the design of the cardiogenic 3D bioprocess was demonstrated.

In conclusion, all the studies reported in this section reveal the ability of CFD-based modelling to obtain reliable and consistent models of cell interactions with the macro- and microenvironment (i.e., scaffolds and bioreactors). Nevertheless, although significant achieved results, future advances in this field are needed. We believe that they will be driven not only by algorithmic innovations allowing more efficient modelling technique, but also by well-characterized and validated experimental data that produces the rules underlying the models.

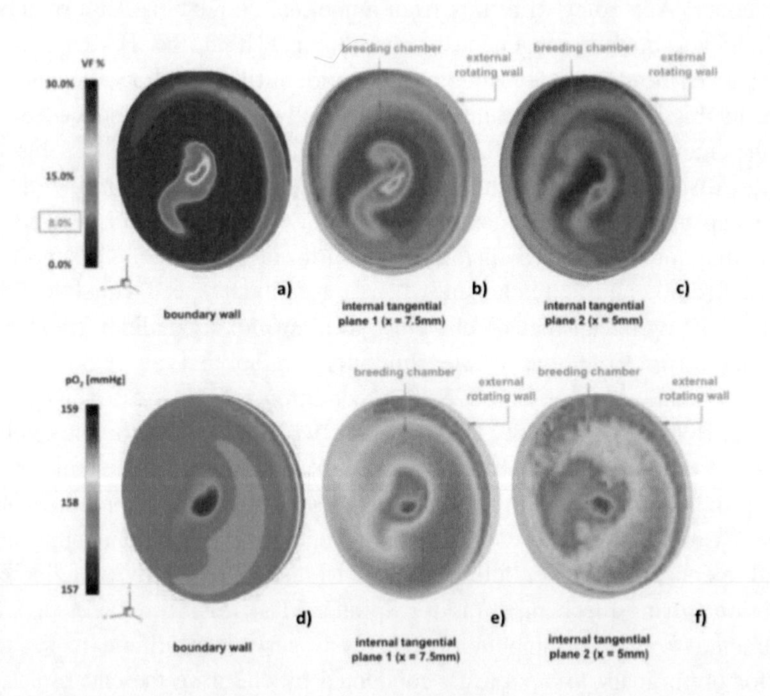

Figure 9. Top panel: contour maps of the volume fraction (VF, expressed in percentage) of the alginate-beads floating within the fluid domain of the 3D model, after 120-s of simulation. Simulation with the 3D model demonstrates that a rotational speed of 25 rpm allows a well-mixed distribution of the secondary phase to be achieved: nor excessively packed (VF_{max} = 30%) neither deposited beads are identifiable. (a) external boundary wall; (b) internal tangential plane at x = 7.5mm; (c) internal tangential plane at x = 5 mm. Bottom panel: contour maps of pO_2 within the fluid domain of the 3D model, after 120-s of simulation; pO_2 is between the range 157 - 159 mmHg in the whole fluid domain, and no zones suffering from insufficient oxygenation are observed. (a) external boundary wall; (b) internal tangential plane at x = 7.5mm; (c) internal tangential plane at x = 5 mm.

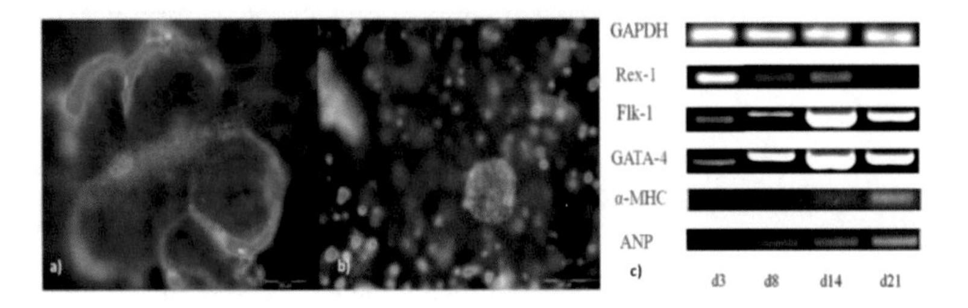

Figure 10. Live/dead assay of cells within alginate beads. mESC-DC viability was assessed at day 21 (a) of cardiac differentiation within HARV bioreactor and (b) after 21 days of static culture carried in petri dish and reported as a control; scale bar 200μm; c) RT-PCR analysis of cardiac gene expression of the mESCs after 21 days of dynamic culture within the rotating HARV.

3.4. MAS Modelling Technique

As discussed in previous sections, CFD allows one to model and evaluate an *in silico* environment that is intended to be applied previously to *in vitro* experiments. However, this approach alone is insufficient to optimally predict the design requirements of a TE experiment. Because cells and tissues are dynamic systems, we require a class of models that is also capable of being dynamic and reactive. Multi agent systems (MAS) can meet this need. In the most general survey, MAS modelling is a computational approach that simulates the interactions of autonomous entities (agents) with each other and their local environment to predict higher-level emergent patterns, through the definition of a rule set governing the actions of each individual agent [Thorne et al. 2007]. Historically, this technique has been widely used in the ecological and social sciences and it has only recently been applied in biomedical research to study complex multi-cell biological phenomena. It is here anticipated that, in TE, MAS modelling technique is aimed at dealing with the attempt to simulate the spatial-temporal dynamics of complexities occurring in *in vitro* cell cultures and how different culture conditions influence the cell life. MAS modelling can therefore be used as a real-time controller for cell culture in bioreactors, beyond as a predictive tool. By means of MAS modelling individual agent (i.e., the cells) dynamics can be explicitly represented in space and time. This modelling approach can provide information about how tissue patterns emerge as a result of cellular interactions within the framework of the tissue-level environment. Furthermore, MAS modelling, which can compute cell behaviours while integrating cell-signalling events, is useful for its ability to facilitate high-throughput *in silico* experiments in a cost- and time-efficient manner.

As anticipated earlier, the biological systems in general and the tissue engineering processes in particular are complex systems. Essentially, four features can describe a complex system [Coveney and Highfield, 1996]:

(i) self-organization: the spontaneous emergence of non-equilibrium structural organisation on a macroscopic level, due to the collective interactions between a larger number of (usually simple) microscopic objects
(ii) non-linearity: changes occur in a non linear fashion
(iii) dynamics: the system is not static, but highly dynamic, changing almost continuously

(iv) emergent properties: properties that occur at higher level and that cannot be predictable from the features exhibited at lower scale

Although so far many biological systems have been described by sets of differential equations, mathematical models lack the plasticity necessary to depict these scenarios. Even simple biological changes can have profound effects on the equations, requiring new equations to be introduced or the modification of many existing equations and many terms in each. It also makes it difficult for model building to scale as systems become increasingly complex. Moreover, differential equation-based models are difficult to reuse when new aspects of cell structure need to be integrated. For these reasons, there is the need of a dynamical computational method that allows zooming back and forth between lower and higher scale behaviours during *in silico* experiments to study emergence and complex phenomena. In that sense, MAS simulation is a promising method to manage that complexity [Russel and Norvig, 1995]. The MAS approach takes advantage of the following properties: i) a modular approach, which makes system integration and maintenance easier and less costly [Sycara, 1998]; ii) collaboration and communication among single agents in a network [Bradshaw et al., 1997]; iii) ability to detect and respond to important time-critical information [Cutkosky et al., 1993]; iv) the possibility to divide problems into a set of more manageable smaller problems [Sycara, 1998]. The underlying philosophy of MAS is that individual agents respond to events and factors in their environment using a set of rules. They then make decisions that lead to a particular behaviour, and the aggregate of agent behaviours over time can produce emergent phenomena [Thorne et al. 2007]. Simulation of interactions of a collection of agents responding to local environmental conditions and neighbouring agents can lead to generation of higher-level patterns. Furthermore, inclusion of multiple agent types allows exploration of mechanistic explanations of more complex behaviours. Drawing on these properties, MAS can, in limited ways, behave similar to tissues, and that may provide insight into how tissue patterns emerge as a consequence of cellular interactions within the framework of the tissue level environment.

Agents represent the basic element of the MAS modelling approach. However, as is often the case for young area of research, there is no universally accepted definition of an agent. As Nwana noted, intelligent agent is now used as an umbrella term to represent a range of software with different characteristics and abilities [Nwana, 1966; see also Franklin and Graesser, 1996]. However, the definition provided by Wooldridge and Jennings is widely accepted "an agent is a computer system situated in some environment that is capable of autonomous action in this environment in order to meet its design objectives" [Wooldridge and Jennings, 1995].

It is clear that the two basic properties of intelligent agents are that i) they are autonomous and that ii) they are situated in an environment. Being autonomous, means that agents are independent and make their own decisions: they have control over its own actions. In that way they are different from active objects. Agents are often proactive: an agent will also be able to act without human intervention. The second property (situatedness) regards the type of environment in which agents exists. Agents tend to be particularly useful when the environment is challenging, i.e. dynamic and unpredictable [Russell and Norvig, 1995; Wooldridge, 2002]. An agent need not assume that the environment will remain static while it tries to achieve a goal. An agent is typically knowledgeable of its local environment but ignorant of components and/or events elsewhere. Relative to a particular agent, such environments can be unreliable. Also, an agent can have multiple persistent goals. Agent actions are influenced by environmental changes as well as its goals. There may be several ways to design agent logic to do both and avoid conflict; there may be no best design. Depending on micro-mechanistic objectives, the logic can be simple and inflexible, or it may be flexible, allowing for multiple options. Agents almost always need to interact with other agents. That means agents are social. In some applications, the significant unit of analysis and design is the individual agent, whereas in other cases it is the society of agents, depending essentially on the problem domain. Agent interaction can be described in terms of human interaction types such as negotiation, coordination, cooperation and teamwork. When the system being represented is particularly complex, large or unpredictable, one can use agents to represent subsystems. A subsystem agent may contain a system of more fine-grained agents. In essence, the problem space is partitioned into a number of smaller and simpler components, which are easier to develop and maintain, and which are specialized at solving the constituent sub-problems. Such decomposition allows each agent to employ the most appropriate paradigm for solving its particular problem, rather than being forced to adopt a common uniform approach that represents a compromise for the entire system.

It is clear from the preceding text how MAS could be used to expedite TE. In TE models based on MAS approach, the key components would be agents representing cells [such as in Bailey et al., 2007; An, 2004; Longo et al., 2004; Meyer-Hermann et al., 2006; Peirce et al., 2004; Segovia-Juarez et al., 2004; Walker et al., 2004; Walker et al., 2004; Grant et al., 2006], although models have used agents at lower levels (i.e. protein systems) and at higher levels (i.e. organs interaction) [Casal et al., 2005]. An agent that is an *in silico* cell can be designed to exhibit a range of biologically relevant behaviours (such as proliferation, apoptosis, migration, differentiation and so on) that are modulated on the basis of the environment. Such modulation can be rendered through a set of condition-action rules that guide agent actions. They may, for example, be based on hypothesized causal factors affecting behaviours. Depending on the targeted mechanisms analyzed, rules can take any form (for instance 'if-then-else' form [Walker et al., 2004] or axioms formalism [Grant et al., 2006]). As the model mature, a rule can be replaced by another MAS, an equation model, or a continuum model. Generally, rules are based on literature or experimental evidence. Rules can vary in sophistication, depending on how agent behaviours are validated and the degree to which each agent event history influences the rules or rule use. When data or experimental techniques are unavailable they can be systematically parameterized [Thorne et al., 2007]. Each rule describes an agent reaction to a set of specific environmental conditions, its neighbour state, and/or its own internal state (fig. 11). Another way to improve the realism of a MAS model is to modify the granularity of the system. Agents can be either atomic or composite. Atomic components define the system level of resolution, i.e., its granularity. Granularity is the extent to which a system is subdivided, with the smallest components being atomic. An atomic object has no internal structure and so cannot be subdivided. Granularity is also the level of specificity or detail with which system content is described: the more fine-grained, the more specific. Objects, both atomic and complex, are pluggable and can be replaced (as distinct from being subdivided) with more fine-grained components that exhibit the same behaviours within the system under the same conditions. That replacement can take place in any time, even during the simulation. Thanking to this feature, the phenomena emerging from mechanisms at one level can be used as input at another level. Greater nesting means more components, and that means more interactions and more simulation time to process, document, and record those interactions. In order to maintain parsimony, models should be designed with components that are just fine-grained enough to produce targeted phenomena and achieve the models specified uses. Also the time-scale of a MAS simulation needs to be tailored to the specific biological hypothesis addressed. Simulations move through time in a series of discrete time steps that can map to any granularity of time, from milliseconds [Casal et al., 2005] to years [Pierce et al., 2004; Abbott et al., 2006].

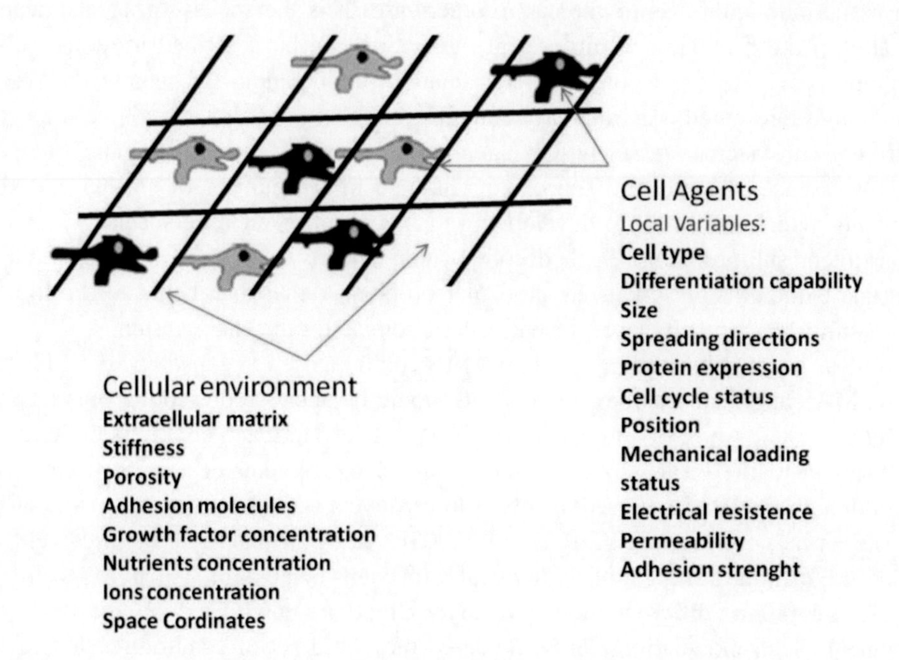

Figure 11. Structure of an agent-based model. Cells are modeled as agents that have a set of local variables that constitute their state. Agents can move around in the cellular environment (that simulate the extracellular matrix), a Cartesian coordinate space, each point of which holds environment variables, such as the local concentration of growth factors, mechanical properties as stiffness and porosity, nutrient and ion concentration. The interactions of different agents with each other and their environment rules are simulated by MAS model according to a set of literature derived.

In developing a MAS simulation, beyond defining the rules governing the agent behaviour, the simulation space needs to be defined. The simulation space represents the cell external environment, which contains relevant non-cellular parameters for the experimental question being addressed. For example, it may include concentrations of diffusible factors (single proteins) [Bailey et al., 2007] or non-diffusible factors (as the extracellular matrix, ECM) [Peirce et al., 2004]. The simulation space in which the agents move can be given closed, open or periodic boundaries. With closed boundaries, agents cannot move past the boundaries, and proteins cannot diffuse beyond boundaries [Longo et al., 2004]. In this case, the simulation area must be sufficiently large relative to the area of agent interactions to avoid edge effects. Conversely, open conditions remove cells or protein from the simulation entirely when they pass beyond the boundaries (i.e., total flux of agents is not a priori set to zero). Periodic (also known as toroidal) conditions are used when modeling a representative section of a larger tissue, wrapping flow across the boundaries to the opposite side so that an agent exiting the simulation through the left-hand border would automatically re-enter the simulation space from the right-hand border [Segovia-Juarez et al., 2004]. Boundary conditions must be carefully chosen depending on the goals of the simulation, and can differ between agents and other quantities within the same space. For example, quantities of proteins may be assumed to approach a background tissue level as they approach the simulation boundaries (Dirichlet or open boundaries), while periodic boundary conditions could be used for agents when modeling a random walk pattern of migration in a portion of a larger tissue.

3.5. MAS Applications in Biomedical Research and Tissue Engineering

Several MAS models have been proposed for the study of cell behaviours. Their use have included discovering plausible mechanisms that enable mimicking some of the complex behaviours occurring *in vivo* in normal and pathological tissues, such as embryogenesis, immunology and cancer biology, producing new understanding with regard to how cells interact with one another and their environment to generate a tissue structure [Keller, 1980; Meyer-Hermann, 2002; Marsden and DeSimone, 2003; Segovia-Juarez et al., 2004; Casal et al., 2005; Meyer-Hermann and Maini, 2005; Meyer-Hermann et al., 2006]. MAS simulations have also been applied to capture interactions between different cell types and atypical gene expression as happens in tumorgenesis and angiogenesis [Abbott et al., 2006; Peirce et al., 2004]. To provide realistic informative insights into the functioning of a complex biological system, models must, at some level, be compared with experimental work. In figure 12 is shown how experiments and MAS can be integrated with one another.

Even if MAS has been applied for several *in vivo* studies, the need to start from clear experimental data to generate the rule sets and to fine tune the parameters during parameterization phase drives MAS applications to model *in vitro* events that generally offers more precisely controlled experimental and simulation parameters and permits in validation and prediction phases to easily design new experimental sets [Gibson et al., 2006; Walker et al., 2004; Walker et al., 2006; Grant et al., 2006]. This simplifies the modeler work during the entire MAS cycle: in fact, inputs, outputs and boundary conditions of both *in vitro* and *in silico* spaces can be explicitly defined. In these ways, MAS of *in vitro* phenomena are more easily paired with wet-lab experiments. Hereafter, we report brief description of nine models that have been developed using a MAS framework and are relevant to either TE or RM scenario. Walker et al. [Walker et al., 2004a; Walker et al., 2004b; Walker et al., 2006a; Walker et al., 2006b] and Sun et al. [Sun et al., 2007; Sun et al., 2008] used a formally based software agent to examine *in vitro* growth and wound-healing mechanisms in urothelium and skin. In their simulations, the transitions between agent states were controlled by functions which were either internal to the agent, or external functions of any level of complexity, allowing, e.g., finite element models of forces between cells, or differential equation models of signalling events, to be linked to the individual-based cell model. Grant et al. [Grant et al., 2006] developed a model to simulate how individual epithelial cells (ECs) organize into multicellular structures.

ECs were studied *in vitro* to help answer that question. Characteristic growth features, which include stable cyst formation in embedded culture, inverted cyst formation in suspension culture, and lumen formation in overlay culture, were investigated under the hypothesis that formation of these characteristic structures is a consequence of an intrinsic program of cellular polarization and depolarization. Bailey et al. [Bailey et al., 2007] integrated signalling events and cell behaviours within a unified spatial and temporal framework at the multi-cell tissue-level. They developed a model of monocyte trafficking in the microvasculature with rules for monocyte rolling dependent on chemokine levels, wall shear stress and adhesion molecule expression.

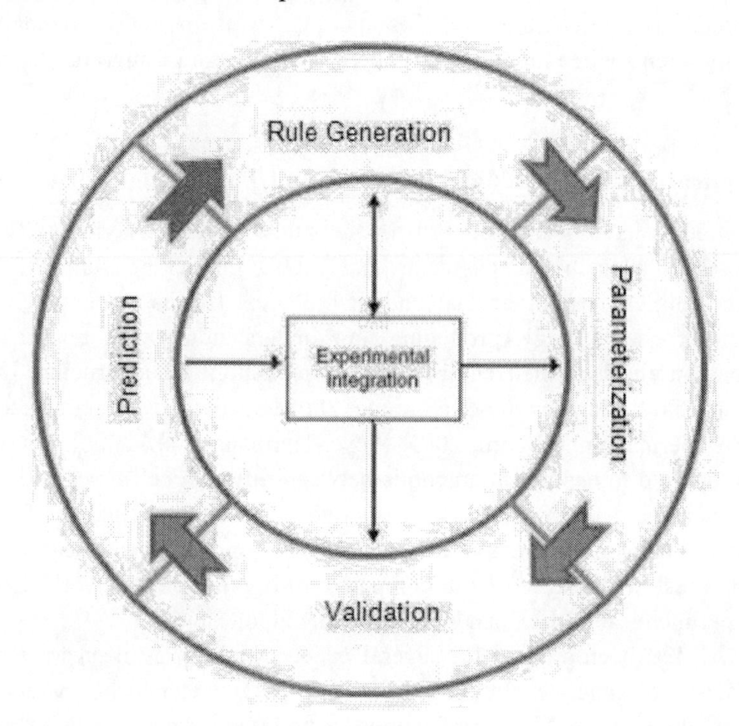

Figure 12. Integration between MAS cycle and laboratory experimentation. Cell interaction experimental data feed into the rule-generation phase, which can stimulate further experimentation if needed data are not available. During the parameterization phase, parameters are varied until model outputs remain within the bounds of a higher-level biological scale experimental data set. After parameterization, models are validated on a different data set. A validated model is used to predict the outcomes of experiments to test novel hypotheses, and helps to design future laboratory experimentation to verify model results. The prediction phase could also lead to expansion of the model, starting another iteration of rule generation, parameterization, validation and prediction. (re-arranged from Thorneet al., 2007).

In vitro experiments allowed control of all the inputs (cell type, adhesion molecule presence, mechanical force presentation and soluble chemokine levels) for the MAS simulation and, when joined with the *in silico* MAS modelling, resulted in quantifiable high-resolution observations, something that is considerably more difficult to achieve *in vivo*. Bindschadler et al. [Bindschadler and McGrath, 2007] presented a model in which a confluent monolayer of cells was injured by forcibly removing a strip of cells. The remaining monolayer healed through a combination of cell migration, spreading and proliferation. Christley et al. [Christley et al., 2007] tried to mimic patterns of mesenchymal condensation in which cells of the embryonic vertebrate limb in high-density culture underwent chondrogenic pattern formation. They also used wet-lab values of diffusion coefficients to specify parameter values involving into the culture. Robertson et al. [Robertson et al., 2007] developed a multiscale model to explore issues of morphogenesis, specifically mesendoderm migration in the Xenopus laevis explant model. Equation based, mass action kinetic models specified intracellular processes. Events at the cell level used an agent-based model of mesendoderm migration across a fibronectin extracellular matrix substrate. Lam et al. [Lam and Hunt, 2009] used MAS for developing a simplified perfused liver model. The goal was discovering plausible mechanistic details of hepatic drug interactions. A similar issue was addressed by Park et al. that described a more detailed, realistic liver model that was used to gain insight into hepatic drug disposition in normal and diseased livers [Park et al., 2010]. Kim et al. [Kim et al., 2009; Kim et al., 2009] achieved a deeper insight into a fundamental feature of *in vitro* morphogenesis through an agent based model of primary human alveolar type II epithelial cells that form alveolar-like cysts using a cytogenesis mechanism that are different from proliferation and death. The model was reused to study wound healing. The results underscored how directional cell locomotion is critical for successful of *in vitro* wound repair rather than proliferation [Kim et al., 2010].

The preceding is a subset of MAS models that have been developed for use in biomedical context. Each one is interesting from a TE point of view because of the model design and the results obtained. Since TE MAS-based modelling is only at the beginning, each work is useful because it represents a starting point to be refined and further extended.

3.6. MAS for in silico Bioreactor

MAS-based modelling in TE is at its early stage of development. In particular, to date, few works have been focused on building *in silico* bioreactor models. One of the most relevant works in this direction is the study by Gaetano et al. [Gaetano et al., 2010] that developed a MAS simulation to support cardiac TE.

According to its definition [Langer and Vacanti, 1993], cardiac TE guides the attempt to repair damaged cardiac tissue through the transplantation of healthy, functional and propagating cells that are capable of restoring tissue viability and function. To achieve that vision, the following needs must be met: an abundantly available, easily accessible, renewable and expandable *in vitro* cell source; a well-defined protocol for driving defined lineages of specific cell types without contamination by undesired elements; and delivery strategies to target required numbers of cells to the target region. Important progress has been made, but cell therapy is still in its infancy. Today, the most important protagonists are stem cells [Segers and Lee, 2008], because of their ability to self renew and differentiate into many different cell types, including cells of the cardiac lineage. Stem cell culture conditions play fundamental roles in such a differentiation process. The properties of cultured tissues, in fact, depend on cultivation conditions, such as type of scaffold, culture media, and bioreactor. For this reason, a problem in cardiac tissue engineering is providing an appropriate environment for the *in vitro* culture. Moreover, it is important to understand how to maintain cells under conditions that maximize their ability to perform their physiological roles. Factors that must be considered include: i) the assessment of relevant cell properties; ii) the measurement and control of *in vitro* culture key parameters; iii) robust behaviour prediction strategies; iv) methods for interrogating and evaluating the many parameters that may impact the culture output; and v) a strategy for testing the many different parameters that may impact cell output in a high throughput and scale-relevant manner. Such studies are extremely time and cost-intensive [Lollini et al., 2006]. By experimenting on MAS we expected to provide a rational method to shrink the problem space, while bringing into focus the more critical tissue engineering issues. In detail, the aim of the work by Gaetano et al. was to design, implement, and initiate validation of a simulation framework capable of integrating information reported in literature, and simulating and predicting stem cell behaviours under specific culture conditions. The key component of the model was the single stem cell that became an agent, which was able to perform the basic processes occurring in a cell life, such as migration, adhesion, growth, division, etc. In the model, individual cells sense their environment and adjust behaviours in response to environmental changes. Biochemical processes occurring inside the cell were conflated so that they could be represented by rules. Each cell was represented using a sphere because isolated cells often adopt a spherical morphology. Assuming that receptors/ligands were homogeneously distributed on the cell surface and the fluctuations of the number of binding sites during the formation and the release of bonds could be neglected, cells in contact could form adhesive bonds. With decreasing distance between cell centres (e.g., upon compression), system density increased exceeding physiological moderate compression resulting in contact inhibition. In anchorage dependent confluent monolayers, as well as in anchorage independent cell spheroids, increasing cell density stopped proliferation: an otherwise normal cell went into a quiescent state. Quiescent cells may undergo density-dependent inhibition of migration or apoptosis. Adhesive bonds between cells and substrate can be considered as resulting from a homogeneous film of ligands. When cells lost this substrate contact, they underwent a form of specific programmed cell death (i.e., anoikis). Depending on cell-cell contacts and/or cell-substrate contacts, a cell can migrate performing a random walk-like movement, or move toward a chemotactic signals. An essential two-phase cell cycle was specified: i) the interphase, in which a cell doubles its mass and its volume; and ii) the mitotic phase, in which a cell divides into two identical daughter cells. The orientation of cell division is set to be stochastic. That is because in this model the polarized structure of the cell cytoskeleton was neglected.

The biological processes above provided specification of features of an *in silico* model, based on principles of MAS design [Russel and Norvig, 1995] coupled with discrete event simulation [Zeigler et al., 2000]. Three discrete objects with eponymous names were defined: CELLS, MATRIX, and FREE SPACE (CAPS was referred to the in silico counterparts). Each CELL was an agent characterized by an inner state. The inner state was composed by: radius (R), number of links between the CELL and its neighbourhood (NL), a percentage representing the CELL attachments to substrate (CSA), CELL doubling time (CDT), and phase (P). Depending on inner state, simulation time, and culture conditions, each CELL tried to achieve its main goals: survival and proliferation along with obeying specific rules. A phenomenological approach was taken: every rule sprang from experimental observations rather than conceptual considerations. The parameters used have been experimentally determined. Each rule took the form of an axiom, which specified a precondition and corresponding action (see Table 3). Preconditions corresponded to a CELL inner state and/or environmental conditions. For any precondition, a corresponding action was specified. A CELL selected just one axiom and completed its corresponding action during each simulation cycle, which mapped to one wet-lab hour. MATRIX and FREE SPACE were passive objects that essentially mapped to units of extracellular matrix (ECM) and matrix-free material. A MATRIX object mapped to a cell-sized volume of ECM or scaffold with appropriate features. A FREE SPACE object mapped to a similarly sized volume of material that was essentially free of cells and matrix elements. For simplicity, FREE SPACE represented any media containing sufficient nutrients, oxygen and so on, unless specified otherwise. A single user interface was designed. It allowed the research to select use modality and specify experimental settings (fig. 13). The researcher also selected CELL type. The researcher either chose to set other variables that influenced the cultures or use default settings. Exploratory experiments were conducted to optimize experimental conditions or to achieve pre-specified outcomes in the least time and at the lowest costs. For this first application, the *in silico* model was implemented using Netlogo 3D (http://ccl.sesp.northwestern.edu/netlogo/), because this system was suitable for exploratory simulations to assess axiom specifications and their relations.

Table 3. An example of axioms that guide CELL behaviours

Preconditions	Actions
P = interphase $NL < NLmax$ $R < 2 \cdot R$	$R = R + dR$ $NL = NL + 1$ (in presence of enough near cell)
P = interphase $NL \geq NLmax$ $R < 2 \cdot R$	P = quiescence state
P = interphase $NL \geq NLmax$ $R = 2 \cdot R$	P = quiescence state
P = interphase $NL < NLmax$ $R = 2 \cdot R$	P = mitotic phase

For the first implementation and initial face validation, we focused on a typical process that occurs in culturing embryonic stem cells (ESCs) in suspension: the formation of embryoid bodies. ESCs, isolated from the inner cell mass of blastocyst stage embryos, are pluripotent cells capable of differentiating into all somatic and germ cell types. There are many ways to allow ESCs to differentiate spontaneously *in vitro*. The basic scenario for inducing differentiation is to form three-dimensional cell aggregates termed embryoid bodies (EBs). Various culture systems have been used for the formation of EBs *in vitro* [Kurosawa, 2007]. These include liquid suspension culture, methyl-cellulose cultures, and hanging drop cultures. It is believed that a non-adherent environment facilitates cell aggregation. In this scenario, with our *in silico* model, we evaluated the importance of geometry of non-adherent culture well on EBs formation. One hundred CELLs, representing ESCs, were seeded into two different well-like objects: one had a cylindrical shape with a round bottom, and the other had the same shape but with a flat bottom. For all the other experimental conditions, parameterizations were default values. A two-day culture was simulated. Each CELL followed the axioms in table 3.

The environment that surrounded CELLs was only FREE SPACE. After seeding, CELLs fell down and settled on the bottom (fig. 14 and 15). The non-adherent wall and the geometry of well influenced sedimentation increasing or not aggregation among CELLS. The first EBs began to form after 6 hours (fig. 14c and 15c). Within the next several hours, a continuum aggregation and proliferation increased EB size (fig. 14d,e and 15d,e). At the end of the simulation, the influence of well geometric characteristics on EBs formation was clear: in a round bottom wells, unique and larger EBs were formed (fig. 16a), whereas in flat bottom wells, more numerous, smaller, and more heterogeneous EBs were generated (fig. 16b).

Cell

Type

Seeding number

Scaffold

Dimension – x y z [mm]

Material

Coating molecules

Medium

Type

Quantity

% of serum

Environment

pH

O_2

CO_2

Temperature

Figure 13. Interface designed for the software tool.

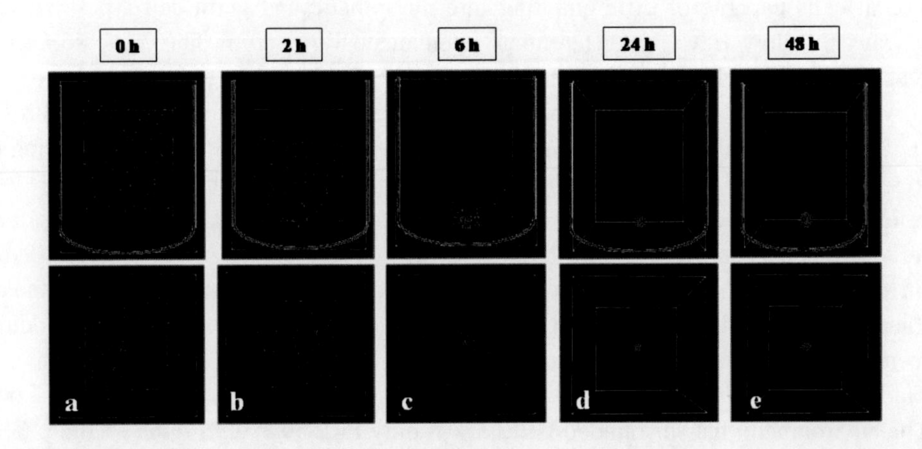

Figure 14. Formation of EB in a round bottom well. First row represents the x-z plane of the well, and the second represents the x-y plane. Results were provided in terms of snapshots of simulated cultures, cell growth curves, and graphs for the time evolution of interesting parameters.

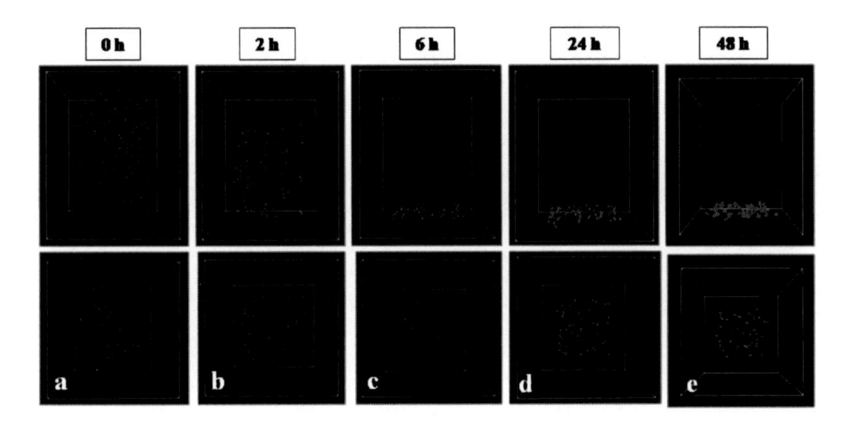

Figure 15. Formation of EB in a flat bottom well. First row represents the x-z plane of the well; the second represents the x-y plane.

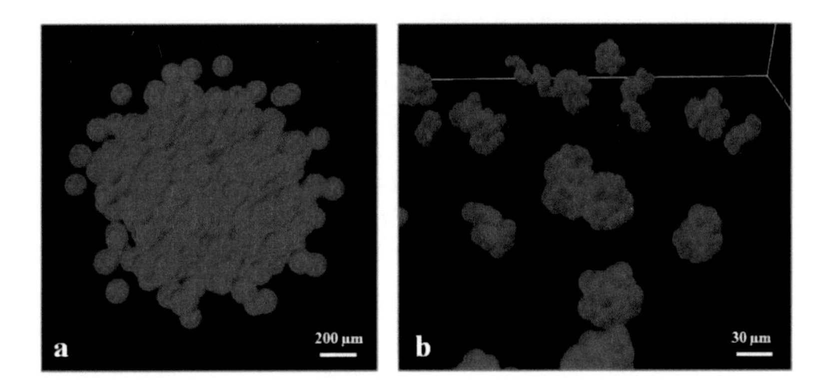

Figure 16. EBs formed after 48 hours in a round bottom (a) and in a flat bottom well (b).

Comparing the dimension of EB formed in round-bottom wells with literature data [Kurosawa, 2007] revealed no differences. Upon evaluating the growth curves, it was clear that there was no significant difference between numbers of cells produced in the two conditions (fig. 17). *In silico*, the geometric condition influenced mainly the spatial arrangement of the CELLs and not their proliferation. Comparison with published data [Koike et al., 2005] revealed no meaningful difference (fig. 17).

In conclusion, using a multi-agent approach, we modelled and simulated the basic biological processes occurring in a specific cell culture. The first set of validation experiments targeted the formation of EBs in suspension culture. Starting with single embryonic stem CELLs, axiom implementation led to results that matched those reported in literature for EBs grown in round or flat-bottom culture wells. The current level of biomimicry supports hypothesizing that there are biological counterparts. Moreover, these preliminary results are supportive of the more long-term goal of using these methods to expedite cardiac tissue engineering. This work is the first step in a process of building and validating biomimetic software systems able to help researchers engaged in cardiac tissue engineering. The next step will deal with the refinement of the model, increasing its complexity and improving its adherence to reality to effectively build an in silico bioreactor.

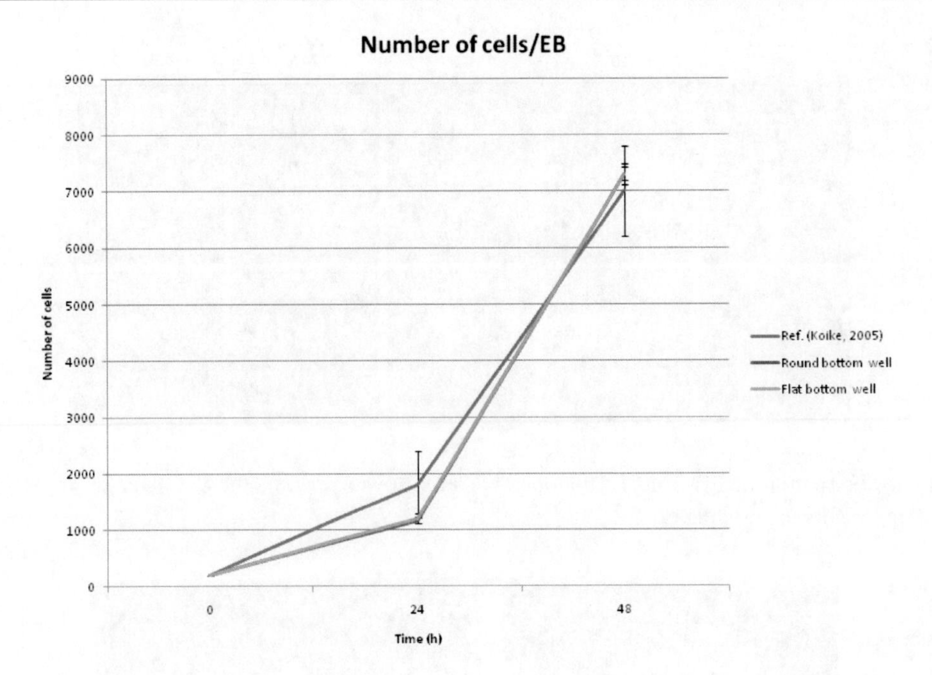

Figure 17. Growth curves for wet-lab (literature referenced), the round bottom, and the flat bottom well condition are compared.

4. CONCLUSION

Within this chapter, recently developed enabling tools assisting and extending traditional experimental TE methodologies have been introduced. In particular, a computational multidisciplinary approach has been described, based on the integration of i) CFD-based simulations, to provide analysis of cell-environment interactions during cell culture in bioreactors and ii) MAS modelling, which can compute and track dynamic change in cell behaviour. The described computational approach has the perspective to become a central tool for tissue engineers, helping to overcome major limitations that currently make impracticable the realization of fully functional bioartificial tissues and organs. The introduction of computational modelling in TE and the employment of research experiments conducted using advanced control strategies will allow to predict conditions for which cells can be optimally cultured, leading to the formation of a physiological-like functional tissue, and the possibility to collect and analyze useful information and data prior to wet-lab experiments. This will allow elucidating those aspects representing the "dark matter" in the experimental setup together with the enormous advantage of saving up time and money.

Unfortunately, despite encouraging achieved results, we are far from having a software tool that could be really intended as an *in silico bioreactor*, i.e., an *in silico* framework able to take into account all (or quite all) the variables involved in TE experiments, and to be used in experimental practice. To *in silico* simulate an *in vitro* tissue culture, in fact, presupposes a model that is able to describe as realistic as possible, and at the same time, all the biological processes occurring during cell culture. Even if this is a very challenging issue, some steps are letting us move towards that goal. The first obstacle arising is the choice of the most suitable numeric tool to developing the model of an *in silico* bioreactor. Here, we presented a traditional approach, equation based (CFD), and a newer one, intelligent agent based (MAS), to address the attempt to depict the multiscale and multiphysic processes occurring in TE experiments. However, to date, combination of these tools and the development of a software framework capable to perform a comprehensive analysis, as the two methods individually do, appears to be impracticable. CFD-based modelling can be used to predict as a cell will respond to different culture conditioning procedures but it is unable to consider the cells as dynamic and reactive entities: in CFD simulations, cells behaviour is predetermined at the beginning of the simulation and cannot be modified during the simulation running. In turn, in MAS modelling, agent behaviour can vary as a function of the defined set of rules according to external environment conditions but, for example, analysis of transport phenomena, i.e., time-dependent analysis of external environment condition variation within the cell breeding chamber, is almost unworkable.

Moreover, as always, each method has advantages and drawbacks. For instance, one of the major reasons for using MAS is that they allow for tracking of individual cells and cell properties. This is something that cannot be done in most *in vivo* experimental settings and is tedious or impossible to do with equation-based models. The heterogeneous properties of tissue are also more easily represented in a MAS model. One of the drawbacks to MAS is that, at commonly used levels of abstraction, some necessary parameters are not concrete: a sort of stochasticity is introduced. However, this stochasticity allows for emergent properties during MAS simulations. On the other hand, at common levels of abstraction for differential equations, parameters often involve reaction rates and binding coefficients, properties that can be definitively measured. However, the deterministic equation-based approach does not permit to properly simulate dynamic cell behaviours.

Another important issue to be addressed in the construction of a real time controller *in silico* bioreactor is the computational time required to perform numerical simulations. The speed of execution of MAS may be more dependent on the skill of the programmer than for equation-based models. That said, larger MAS can monopolize computational resources, and few MAS suites allow for parallel processing to reduce computational time. Anyway, in both cases, a large number of simulations are required for adequate exploration of the parameter space. In equation-based models, there are well-developed mathematical methods for this exploration, while these techniques are just beginning to be applied to MAS. Finally, MAS require the assumption that all properties can be modelled discretely, while the other approach require a continuum approximation. Technically, this represents the most difficult aspect precluding intuitive communication between CFD and MAS models.

Maybe, the reality for biological systems may lie somewhere in the middle of these and other approaches. For this reason, new approaches are being developed to bridge MAS and CFD modelling techniques [Peirce et al., 2006] in order to capitalize on beneficial attributes of these approaches and to compensate for their respective drawbacks. We are working on coupling these two methods to join cells, described with MAS, with their environment, composed by fluids, nutrients, and mechanical stresses and rendered by CFD.

In conclusion, the need for computational simulation and prediction of multi-cellular events in space and time becomes apparent [An, 2005; Hunt et al., 2009]. Each advantage of each approach must be further exploited in name of the next future. Following this perspective, the ultimate aim of computational modelling in TE should be the development of a tool that, starting from the acquisition of patient-specific data, is able to guide researchers in designing and performing a cell culture experiment resulting in the formation of an engineered functioning portion of tissue to be *in vivo* implanted.

REFERENCES

Abbott, R., Forrest, S., Pienta, K. Simulating the hallmarks of cancer. *Artif Life*, 2006, 12, 617–634.

Altman, G.H., Lu, H.H., Horan, R.L., Calabro, T., Ryder, D.; Kaplan, D.L., Stark, P., Martin, I., Richmond, J.C., and Vunjak-Novakovic, G. Advanced bioreactor with controlled application of multi-dimensional strain for tissue engineering. *J. Biomech. Eng.* 2002;124, 742.

An, G. In silico experiments of existing and hypothetical cytokine-directed clinical trials using agent-based modeling. *Crit Care Med*, 2004, 32, 2050–2060.

An, G. Mathematical modeling in medicine: a means, not an end. Crit Care Med, 2005, 33, 253-254.

Andreadis, S.T. Gene-modified tissue-engineered skin: the next generation of skin substitutes. *Adv Biochem Eng Biotechnol*, 2007, 103, 241-274.

Atala, A.; Bauer, S.B.; Soker, S.; Yoo, J.J.; Retik, A.B. Tissue-engineered autologous bladders for patients needing cystoplasty. *Lancet*, 2006, 367, 1215-1216.

Bailey, A.M.; Thorne, B.C.; Peirce, S.M. Multi-cell agent-based simulation of the microvasculature to study the dynamics of circulating inflammatory cell trafficking. *Ann Biomed Eng*, 2007, 35, 916–936.

Ballyns, J.J.; Wright; T.M., Bonassar, L.J. Effect of media mixing on ECM assembly and mechanical properties of anatomically-shaped tissue engineered meniscus. *Biomaterials*. 2010 Sep;31(26):6756-63.

Barzilla, J.E.; McKenney, A.S.; Cowan, A.E.; Durst, C.A.; Grande-Allen K.J. Design and Validation of a Novel Splashing Bioreactor System for Use in Mitral Valve Organ Culture. *Ann Biomed Eng*. 2010 Jul 27. [Epub ahead of print].

Bilodeau, K. and Mantovani, D. Bioreactors for tissue engineering: focus on mechanical constraints. A comparative review. *Tissue Eng*, 2006, 12(8), 2367-2383.

Bindschadler, M. and McGrath, J.L. Sheet migration by wounded monolayers as an emergent property of single-cell dynamics. *J Cell Sci*, 2007, 120, 876–884.

Boschetti, F.; Raimondi, M.T.; Migliavacca, F.; Dubini, G. Prediction of the micro-fluid dynamic environment imposed to three-dimensional engineered cell systems in bioreactors. *J Biomech*, 2006, 39(3), 418-425.

Botchwey, E.A.; Pollack, S.R.; El-Amin, S.; Levine, E.M.; Tuan, R.S.; Laurencin, C.T. Human osteoblast-like cells in three-dimensional culture with fluid flow. *Biorheology*, 2003, 40(1-3), 299-306.

Braccini, A;, Wendt, D.; Jaquiery, C.; Jakob, M.; Heberer, M.; Kenins, L.; Wodnar-Filipowicz, A.; Quarto R., Martin, I. Three-Dimensional Perfusion Culture of Human Bone Marrow Cells and Generation of Osteoinductive Grafts. *Stem cells*; 2005. 23(8): 1066-1072.

Bradshaw, J.M.; Dutfield, S.; Benoit, P.; Woolley, J.D. KAoS: Toward an industrial-strength open agent architecture. *Software agents*, 1997, 1, 375–418.

Breen, L.T.; McHugh, P.E.; Murphy, B.P. Multi-axial mechanical stimulation of HUVECs demonstrates that combined loading is not equivalent to the superposition of individual wall shear stress and tensile hoop stress components. *J Biomech Eng*. 2009 Aug;131(8):081001.

Bursac, N.; Papadaki, M.; White, J.A.; Eisenberg, S.R.; Vunjak-Novakovic, G.; Freed, L.E. Cultivation in rotating bioreactors promotes maintenance of cardiac myocyte electrophysiology and molecular properties. *Tissue Eng*. 2003;9:1243.

Cantini, M.; Fiore, G.B.; Redaelli, A.; Soncini, M. Numerical fluid-dynamic optimization of microchannel-provided porous scaffolds for the co-culture of adherent and non-adherent cells. *Tissue Eng Part A*, 2009, 15(3), 615-623.

Carrier, R.L.; Rupnick, M.; Langer, R.; Schoen, F.J.; Freed, L.E.; Vunjak-Novakovic, G. Effects of oxygen on engineered cardiac muscle. *Biotechnol Bioeng*, 2002, 78, 617-625.

Casal, A.; Sumen, C.; Reddy, T.E.; Alber, M.S.; Lee P.P. Agent-based modeling of the context dependency in T cell recognition. *JTheorBiol,* 2005, 236, 376–391.

Catapano, G. Mass transfer limitations to the performance of membrane bioartificial liver support device. *Int J Artif Organs*, 1996, 19, 18-35.

Christley, S.; Alber, M.S.; Newman, S.A. Patterns of mesenchymal condensation in a multiscale, discrete stochastic model. *PLoS Comput Biol*, 2007, 3(4), e76.

Cioffi, M.; Boschetti, F.; Raimondi, M.T.; Dubini, G. Modeling evaluation of the fluid-dynamic microenvironment in tissue-engineered constructs: a micro-CT based model. *Biotechnol Bioeng*, 2006, 93(3), 500-510.

Cioffi, M.; Kuffer, J.; Strobel, S.; Dubini, G.; Martin, I.; Wendt, D. Computational evaluation of oxygen and shear stress distributions in 3D perfusion culture systems: Macro-scale and micro-structured models. *Journal of Biomechanics*, 2008, 41(14), 2918-2925.

Colton, C.K. Implantable biohybrid artificial organs. *Cell Transplant*, 1995, 4(4), 415-436.

Consolo, F.; Fiore, G.B.; Truscello, S.; Caronna, M.; Morbiducci, U.; Montevecchi, F.M.; Redaelli, A. A computational model for the optimization of transport phenomena in a rotating hollow-fiber bioreactor for artificial liver. *Tissue Eng Part C Methods*, 2009, 15(1), 41-55.

Consolo, F.; Mastrangelo, F.; Ciardelli, G.; Montevecchi, F.M.; Morbiducci, U.; Sassi, M.; Bignardi, C. *Multilevel experimental and modelling techniques for bioartificial scaffolds and matrices*. In: Bhushan B. Scanning Probe Microscopy in Nanoscience and Nanotechnology. Heidelberg: Springer-Verlag; 2010; 425-486.

Consolo, F.; Bariani, C.; Morbiducci, U.; Mantalaris, A.; Redaelli, A; Montevecchi, F.M. Multiscale computational model for the design of a scalable, automated and functional cardiogenic 3D bioprocess ofencapsulated embryonic stem cells. *Proceedings of Tissue Engineering and Regenerative Medicine International Society (TERMIS) - EU Meeting - 2010 Galway*, Ireland.

Couet, F.; Mantovani, D. How to optimize maturation in a bioreactor for vascular tissue engineering: focus on a decision algorithm for experimental planning. *Ann Biomed Eng,* 2010, 2877-2884.

Coveney, P.V. and Highfield, R.R. *Frontiers of complexity*. London: Faber and Faber; 1996.

Cutkosky, M.R.; Engelmore, R.S.; Fikes, R.E.; Gene-sereth, M.R.; Gruber, T.R.; Mark, W.S; Tenenbaum, JM; - Weber, JC. PACT: An experiment in integrating concurrent engineering systems. *Computer*, 1993, 26, 28–37.

Dekel, B.; Burakova, T.; Arditti, F.D.; Reich-Zeliger, S.; Milstein, O.; Aviel-Ronen, S.; Rechavi, G.; Friedman, N.; Kaminski, N.; Passwell, J.H.; Reisner, Y. Human and porcine early kidney precursors as a newsource for transplantation. *Nature Med,* 2003, 9, 53–60.

Dermenoudis, S.; Missirlis, Y. Design of a novel rotating wall bioreactor for the in vitro simulation of the mechanical environment of the endothelial function. *J Biomech*. 2010 May 7;43(7):1426-31.

Doroski, D.M.; Levenston, M.E.; Temenoff, J.S. Cyclic Tensile Culture Promotes Fibroblastic Differentiation of Marrow Stromal Cells Encapsulated in Poly(Ethylene Glycol)-Based Hydrogels. *Tissue Eng Part A*. 2010 Jul 28. [Epub ahead of print].

Du, D.; Furukawa, K.S.; Ushida, T. 3D culture of osteoblast-like cells by unidirectional or oscillatory flow for bone tissue engineering. *Biotechnol Bioeng*. 2009 Apr 15;102(6):1670-8.

Durst, C.A. and Jane Grande-Allen K. Design and physical characterization of a synchronous multivalve aortic valve culture system. *Ann Biomed Eng*. 2010 Feb;38(2):319-25.

Elder, B.D. and Athanasiou, K.A. Hydrostatic pressure in articular cartilage tissue engineering: from chondrocytes to tissue regeneration. *Tissue Eng Part B Rev*, 2009, 15(1), 43-53.

Eschenhagen, T.; Didie, M.; Heubach, J.; Ravens, U.; Zimmermann, W.H. Cardiac tissue engineering. *Transplant Immunol*, 2002, 9, 315–321.

Ferziger, F.H. and Peric, M. *Computational Methods of fluid dynamics*. Berlin: Springer; 1996.

Fink, C.; Ergun, S.; Kralisch, D.; Remmers, U.; Weil, J.; Eschenhagen, T. Chronic stretch of engineered heart tissue induces hypertrophy and functional improvement. *FASEB J.* 14, 669, 2000.

Franklin, S. and Graesser, A. Is it an agent, or just a program?. *Intelligent agents*, 1996, 1193, 21-36.

Freed, L.E.; Langer, R.; Martin, I.; Pellis, N.R.; Vunjak- Novakovic, G. Tissue engineering of cartilage in space. *Proc. Natl. Acad. Sci. USA* 94, 13885, 1997.

Gaetano, L.; Di Benedetto, G.; Morbiducci, U.; Montevecchi, F.M.; Hunt, C.A. An Agent-based System Model to Support Cardiac Tissue Engineering: Preliminary Results. *Proceedings of 19th International Conference on Software Engineering and Data Engineering, 2010.*

Galbusera, F.; Cioffi, M.; Raimodi, M.T.; Pietrabissa, R. Computational modelling of combined cell population dynamics and oxygen transport in engineered tissue subject to interstitial perfusion. *Comput Methods Biomech Biomed Engin*, 2007, 10, 279-287.

Gassmann, M.; Fandrey, J.; Bichet, S.; Wartenberg, M.; Marti, H.H.; Bauer, C.; Wenger, R.H.; Acker, H. Oxygen supply and oxygen-dependent gene expression in differentiating embryonic stem cells. *Proc Natl Acad Sci U S A,* 1996, 93(7), 2867-2872.

Gemmiti, C.V. and Guldberg, R.E. Shear stress magnitude and duration modulates matrix composition and tensile mechanical properties in engineered cartilaginous tissue. *Biotechnol Bioeng*. 2009 Nov 1;104(4):809-20.

Gibson, M.; Patel, A.; Nagpal, R.; Perrimon, N. The emergence of geometric order in proliferating metazoan epithelia. *Nature*, 2006, 442, 1038–1041.

Gomez, C. An unit cell based multi-scale modeling and design approach for tissue engineered scaffolds [online]. 2007. Available from: http://idea.library.drexel.edu/handle/1860/1766

Gooch, K.J.; Kwon, J.H.; Blunk, T.; Langer, R.; Freed L.E.; Vunjak-Novakovic, G. Effects of mixing intensity on tissue-engineered cartilage. *Biotechnol. Bioeng.* 2001;72:402.

Grant, M.; Mostov, K.; Tlsty, T.; Hunt, C.A. Simulating properties of in vitro epithelial cell morphogenesis. *PLoS Comput Biol*, 2006, 2(10), e129.

Hoerstrup, S.P.; Sodian, R.; Daebritz, S.; Wang, J.; Bacha, E.A.; Martin, D.P.; Moran, A.M.; Guleserian, K.J.; Sperling, J.S.; Kaushal, S.; Vacanti, J.P.; Schoen, F.J.; Mayer, J.E. Jr. Functional living trileaflet heart valves grown in vitro. *Circulation*, 2000, 102, III44-49.

Hoerstrup, S.P.; Sodian, R.; Sperling, J.S.; Vacanti, J.P.; Mayer, J.E. Jr. New pulsatile bioreactor for in vitro formation of tissue engineered heart valves. *Tissue Eng,* 2000, 6, 75-79.

Horner, M.; Miller, W.M.; Ottino, J.M.; Papoutsakis, E.T. Transport in a grooved perfusion flat-bed bioreactor for cell therapy applications. *Biotechnol Prog*, 1998, 14, 689-698.

Hunt, C.A.; Ropella, G.; Lam, T.; Tang, J.; Kim, S.H.; Engelberg, J. At the biological modeling and simulation frontier. *Pharm Res*, 2009, 26, 2369-2400.

Ignatius, A.; Peraus, M.; Schorlemmer, S.; Augat, P.; Burger, W.; Leyen, S.; Claes, L. Osseointegration of alumina with a bioactive coating under load-bearing and unloaded conditions. Biomaterials, 2005, 26(15), 2325-2332.

Ikada, Y. Challenges in tissue engineering. *J R Soc Interface*, 2006, 3(10), 589-601.

Ikada, Y. *Tissue Engineering: Fundamentals and Applications*. San Diego: Academic Press; 2006.

Jagodzinski, M.; Breitbart, A.; Wehmeier, M.; Hesse, E.; Haasper, C.; Krettek, C.; Zeichen, J.; Hankemeier S. Influence of perfusion and cyclic compression on proliferation and differentiation of bone marrow stromal cells in 3-dimensional culture. *J Biomech.* 2008;41(9):1885-91.

Jenkins, D.D.; Yang, G.P.; Lorenz, H.P.; Longaker, M.T.; Sylvester, K.G. Tissue engineering and regenerative medicine. *Clin Plast Surg,* 2003, 30(4), 581-588.

Jung, Y.; Kim, S.H.; Kim, Y.H.; Kim, S.H. The effects of dynamic and three-dimensional environments on chondrogenic differentiation of bone marrow stromal cells. *Biomed Mater*, 2009, 25, 4

Kasper, C.; van Griensven, M.; Portner, R. Bioreactor systems for tissue engineering. In: Scheper T. *Advances in Biochemical Engineering/Biotechnology*. Heidelberg: Springer; 2009.

Keller, R. The cellular basis of epiboly: an SEM study of deep-cell rearrangement during gastrulation in Xenopus laevis. *J Embryol ExpMorphol*, 1980, 60, 201–234.

Kelm, J.M.; Lorber, V.; Snedeker, J.G.; Schmidt, D.; Broggini-Tenzer, A.; Weisstanner, M.; Odermatt, B.; Mol, A.; Zünd, G.; Hoerstrup, S.P. A novel concept for scaffold-free vessel tissue engineering: self-assembly of microtissue building blocks. *J Biotechnol.* 2010 Jul 1;148(1):46-55.

Kim, S.H.; Sheikh-Bahaei, S.; Hunt, C.A. Multi-Agent Simulation of Self-Organizing Behaviors of Alveolar Cells In Vitro. *The International Journal of Intelligent Controls and Systems*, 2009, 14(1), 41-50.

Kim, S.H.; Yu, W.; Mostov, K.; Matthay, M.A.; Hunt, C.A. A computational approach to understand in vitro alveolar morphogenesis. *PLoS One*, 2009, 4(3), e4819.

Kim, S.H.; Matthay, M.A.; Mostov, K.; Hunt, C.A. Mar 17. Simulation of lung alveolar epithelial wound healing in vitro. *J R Soc Interface*, 2010, 7(49), 1157-1170.

Kofidis, T.; Lenz, A.; Boublik, J.; Akhyari, P.; Wachsmann, B.; Mueller-Stahl, K.; Hofmann, M.; Haverich, A. Pulsatile perfusion and cardiomyocyte viability in a solid three-dimensional matrix. *Biomaterials*, 2003, 24, 5009-5014.

Koike, M.; Kurosawa, H.; Amano, Y. A Roundbottom 96-well polystyrene plate coated with 2-methacryloyloxyethyl phosphorylcholine as an effective tool for embryoid body formation. *Cytotechnology*, 2005, 47, 3-10.

Kurosawa, H. Methods for inducing embryoid body formation: in vitro differentiation system of embryonic stem cells. *J Biosci Bioeng*, 2007, 3, 389-398.

Kurpinski, K. and Li, S.; Mechanical stimulation of stem cells using cyclic uniaxial strain. *J Vis Exp*. 2007;(6):242.

Laganà, K.; Moretti, M.; Dubini, G.; Raimondi M.T. A new bioreactor for the controlled application of complex mechanical stimuli for cartilage tissue engineering. *Proc Inst Mech Eng H*. 2008 Jul;222(5):705-15.

Lam, T.N. and Hunt, C.A. Discovering plausible mechanistic details of hepatic drug interactions. *Drug Metab Dispos*, 2009, 37(1), 237– 246.

Langer, R. and Vacanti, J.P. Tissue engineering. *Science*, 1993, 260, 920-926.

Lee, J.; Guarino, V.; Gloria, A.; Ambrosio, L.; Tae, G.; Kim, Y.H.; Jung, Y.,; Kim, S.H.; Kim, S.H. Regeneration of Achilles' tendon: the role of dynamic stimulation for enhanced cell proliferation and mechanical properties. *J Biomater Sci Polym Ed*. 2010;21(8-9):1173-90.

Lollini, P,; Motta, S,; Pappalardo, F. Discovery of cancer vaccination protocols with a genetic algorithm driving an agent based simulator. *BMC Bioinformatics*, 2006, 7, 352-361.

Longo, D.; Peirce, S.M.; Skalak, T.C.; Davidson, L.; Marsden, M.; Dzamba, B.; DeSimone D.W. Multicellular computer simulation of morphogenesis: blastocoel roof thinning and matrix assembly in Xenopus laevis. *Dev Biol,* 2004, 271, 210–222.

Lyons, E. and Pandit, A. Design of bioreactors for cardiovascular applications. Topics in Tissue Engineering. *Ashammakhi and Reis*; 2005.

Ma, C.Y.J.; Kumar, R.; Xu, X.Y.; Mantalaris, A. A combined fluid dynamics, mass transport and cell growth model for a three-dimensional perfused bioreactor for tissue engineering of haematopoietic cells. *Biochem Eng J,* 2007, 35, 1-11.

MacArthur, B.D. and Oreffo, R.O.C. Bridging the gap. *Nature*, 2005, 433, 19.

Malda, J.; van Blitterswijk, C.A.; van Geffen, M.; Martens, D.E.; Tramper, J.; Riesle, J. Low oxygen tension stimulates the redifferentiation of dedifferentiated adult human nasal chondrocytes. *Osteoarthritis Cartilage*, 2004, 12(4), 306-313.

Malek, A.M.; Alper, S.L.; Izumo, S. Hemodynamic shear stress and its role in atherosclerosis. *JAMA*, 1999, 282, 2035-2042.

Mareels, G.; Poyck, P.P.; Eloot, S.; Chamuleau, R.A.; Verdonck, P.R. Three-dimensional numerical modelling and computational fluid dynamics simulations to analyze and improve oxygen availability in the AMC bioartificial liver. *Ann Biomed Eng*, 2006, 34, 1729-1744.

Marsden, M, and DeSimone, D. Integrin-ECM interactions regulate cadherin-dependent cell adhesion and are required for convergent extension in Xenopus. *Curr Biol*, 2003, 13, 1182–1191.

Martin, I.; Wendt, D.; Heberer, M. The role of bioreactors in tissue engineering. *Rev Trends Biotechnol*, 2004, 22(2), 80-86.

Martin Y., and Vermette P. Bioreactors for tissue mass culture: design, characterization, and recent advances. *Biomaterials*, 2005, 26(35), 7481-7503.

Meyer-Hermann, M.E. and Maini, P. Cutting edge: back to 'one-way' germinal centers. *JImmunol*, 2005, 174, 2489–2493.

Meyer-Hermann, M.E. A mathematical model for the germinal center morphology and affinity maturation. *JTheor Biol*, 2002, 216, 273–300.

Meyer-Hermann, M.E; Maini, P.; Iber, D.. An analysis of B cell selection mechanisms in germinal centers. *Math Med Biol*, 2006, 23, 255–277.

Miller, WM. Bioreactor design considerations for cell therapies and tissue engineering. *Proceedings of WTEC Workshop on Tissue Engineering Research*, 2000.

Nerem, R.M. and Seliktar, D. Vascular tissue engineering. *Annu Rev Biomed Eng,* 2001, 3, 225-243.

Niklason, L.E.; Gao, J.; Abbott, W.M.; Hirschi, K.K.; Houser, S.; Marini, R.; Langer, R. Functional arteries grown in vitro. *Science*, 1999, 284, 489-493.

Nwana, H.S. Software Agents: An Overview. *Knowledge Engineering Review*, 1966, 11(3), 1-40.

Olivares, A.L.; Marsal, E.; Planell, J.A.; Lacroix, D. Finite element study of scaffold architecture design and culture conditions for tissue engineering. *Biomaterials*, 2009, 30(30), 6142-6149.

Pancrazio, J.J.; Wang, F.; Kelley, C.A. Enabling tools for tissue engineering. Review. *Biosensors and Bioelectronics*, 2007, 22, 2803-2811.

Park, J.; Berthiaume, F.; Toner, M.; Martin, L.; Tilles, A.W. Microfabricated grooved substrates as platforms for bioartificial liver reactors. *Biotechnol Bioeng*, 2005, 90, 632-644.

Park., S; Kim, S.H.; Ropella, G.E.; Roberts, M.S.; Hunt, C.A. Tracing Multiscale Mechanisms of Drug Disposition in Normal and Diseased Livers. *J Pharmacol Exp Ther*, 2010, 334(1), 124-136.

Partap, S.; Plunkett, N.A.; Kelly, D.J.; O'Brien, F.J.; Stimulation of osteoblasts using rest periods during bioreactor culture on collagen-glycosaminoglycan scaffolds. *J Mater Sci Mater Med*. 2009 Dec 20. [Epub ahead of print].

Peirce, S.M.; Van Gieson, E.J.; Skalak, T.C. Multicellular simulation predicts microvascular patterning and in silico tissue assembly. *FASEB J,* 2004, 18, 731–733.

Peirce, S.M.; Skalak, T.C.; Papin, J. Multiscale biosystems integration: coupling intracellular network analysis with tissue-patterning simulations. *IBM J Res Dev*, 2006, 50, 601–615.

Piccinini, E.; Wendt, D.; Martin, I. Bioreactor systems in regenerative medicine: from basic science to biomanufacturing. Proc of Mater Res Soc Symp, 2009.

Pierre, J. and Oddou, C. Engineered bone culture in a perfusion bioreactor: a 2D computational study of stationary mass and momentum transport. *Comput Methods Biomech Biomed Engin*, 2007, 10(6), 429-438.

Pollack, S.R.; Meaney, D.F.; Levine, E.M.; Litt, M.; Johnston, E.D. Numerical model and experimental validation of microcarrier motion in a rotating bioreactor. *Tissue Eng*, 2000, 6(5), 519-530.

Porter, B.D.; Lin, A.S.; Peister, A.; Hutmacher, D.; Guldberg, R.E. Noninvasive image analysis of 3D construct mineralization in a perfusion bioreactor. *Biomaterials*, 2007, 28(15), 2525-2533.

Rabkin, E. and Schoen, F.J. Cardiovascular tissue engineering. *Cardiovascular Pathology*, 2002, 11, 305-317.

Raimondi, M.T.; Moretti, M.; Cioffi, M.; Giordano, C.; Boschetti, F.; Lagana, K.; Pietrabissa, R. The effect of hydrodynamic shear on 3D engineered chondrocyte systems subject to direct perfusion. *Biorheology*, 2006, 43, 215-222.

Ritchie, A.C.; Wijaya, S.; Ong, W.F.; Zhong, S.P.; Chian, K.S.; Dependence of alignment direction on magnitude of strain in esophageal smooth muscle cells. *Biotechnol Bioeng*. 2009 Apr 15;102(6):1703-11.

Robertson, S.H.; Smith, C.K.; Langhans, A.L.; McLinden, S.E.; Oberhardt, M.A.; Jakab, K.R.; Dzamba, B.; De Simone, D.W.; Papin, J.A.; Peirce, S.M. Multiscale computational analysis of Xenopus laevis morphogenesis reveals key insights of systems-level behavior. *BMC Syst Biol,* 2007, 1, 46.

Russel, S. and Norvig, P. Artificial Intelligence. A moder approach. New Jersey: Prentice-Hall; 1995.

Saim, A.B.; Cao, Y.; Weng, Y.; Chang, C.N.; Vacanti, M.A.; Vacanti, C.A.; Eavey, R.D. Engineering autogenous cartilage in the shape of a helix using an injectable hydrogel scaffold. *Laryngoscope*, 2000, 110, 1694-1697.

Saito, A.; Aung, T.; Sekiguchi, K.; Sato, Y.; Vu, D.M.; Inagaki, M.; Kanai, G.; Tanaka, R.; Suzuki, H.; Kakuta, T. Present status and perspectives of bioartificial kidneys. *J Artif Organs*, 2006, 9(3), 130-135.

Schulz, R.M.; Wustneck, N.; van Donkelaar, C.C.; Shelton, J.C.; Bader, A. Development and Validation of a Novel Bioreactor System for Load- and Perfusion-Controlled Tissue Engineering of Chondrocyte-Constructs. *Biotechnology and Bioengineering*, 2008. Vol. 101(4): 714-728.

Segers, V. and Lee, R. Stem-cell therapy for cardiac disease. *Nature*, 2008, 451, 937-942.

Segovia-Juarez, J.L.; Ganguli, S.; Kirschner, D. Identifying control mechanisms of granuloma formation during M. tuberculosis infection using an agent-based model. *J Theor Biol*, 2004, 231, 357–376.

Seidel, J.O., Pei, M., Gray, M.L., Langer, R., Freed, L.E., and Vunjak-Novakovic, G. Long-term culture of tissue engineered cartilage in a perfused chamber with mechanical stimulation. *Biorheology*, 2004; 41, 445.

Semple, J.L.; Woolridge, N.; Lumsden, C.J. In vitro, in vivo, in silico: computational systems in tissue engineering and regenerative medicine. *Tissue Eng*, 2005, 11(3-4), 341-356.

Sengers, B.G.; van Donkelaar, C.C.; Oomens, C.W.; Baaijens, F.P. Computational study of culture conditions and nutrient supply in cartilage tissue engineering. Biotechnology Progress, 2005, 21, 1252-1261.

Shachar, M. and Cohen, S. Cardiac tissue engineering, ex-vivo: design principles in biomaterials and bioreactors. *Heart Fail Rev*, 2003, 8(3), 271-276.

Singh, H.; Teoh, S.H.; Low, H.T.; Hutmacher, D.W. Flow modelling within a scaffold under the influence of uni-axial and bi-axial bioreactor rotation. *J Biotechnol*, 2005, 119, 181-196.

Solanki, A.; Kim, J.D.; Lee, K.B. Nanotechnology for regenerative medicine: nanomaterials for stem cell imaging. *Nanomed*, 2008, 3(4), 567-578.

Spaulding, G.F.; Jessup, J.M.; Goodwin, T.J. Advances in cellular construction. *J Cell Biochem*, 1993, 51, 249-251.

Stoltz, J.F.; Bensoussan, D.; Decot, V.; Ciree, A.; Netter, P.; Gillet, P. Cell and tissue engineering and clinical applications: an overview. *Biomed Mater Eng,* 2006, 16(4 Suppl), S3-S18.

Sucosky, P. Flow characterization and modelling of cartilage development in a spinner-flask bioreactor [online]. 2005. Available from: http://smartech.gatech.edu/handle/1853/6875.

Sun, W. and Lal, P. Recent development on computer aided tissue engineering - a review. *Computer Methods and Programs in Biomedicine*, 2002, 67, 85-103.

Sun, T.; McMinn, P.; Coakley, S.; Holcombe, M.; Smallwood, R.; Macneil, S. An integrated systems biology approach to understanding the rules of keratinocyte colony formation. *J Roy Soc Interface*, 2007, 4, 1077–1092.

Sun, T.; McMinn, P.; Holcombe, M.; Smallwood, R.; Macneil, S. Agent based modeling helps in understanding the rules by which fibroblasts support keratinocyte colony formation. *PLoS One*, 2008, 3(5), e2129.

Sycara, K. Multiagent systems. Artificial Intelligence Magazine, 1998, 19, 79–92.

Thorne, B.; Bailey, A.; Peirce., S. Combining experiments with multi-cell agent-based modeling to study biological tissue patterning. *Brief Bioinform*, 2007, 8(4), 245-257.

Tzanakakis, E.S.; Hess, D.J.; Sielaff, T.D.; Hu, W.S. Extracorporeal tissue engineered liver-assist devices. *Annu Rev Biomed Eng*, 2000, 2, 607-632.

Versteeg, H. and Malalasekra, W. *An Introduction to Computational Fluid Dynamics: The Finite Volume Method Approach*. London: Pearson Education; 1995.

Vismara, R.; Soncini, M.; Talò, G.; Dainese, L.; Guarino, A.; Redaelli, A.; Fiore, G.B. A bioreactor with compliance monitoring for heart valve grafts. *Ann Biomed Eng.* 2010 Jan;38(1):100-8.

Walker, D.C.; Hill, G.; Wood, S.M.; Smallwood, R.H.; Southgate, J. Agent-based modelling of wounded epithelial cell monolayers. *IEEE Trans Nanobioscience*, 2004a, 3, 153–163.

Walker, D.C.; Southgate, J.S.; Hill, G.; Holcombe, M.; Hose, D.R.; Wood, S.M; Mac Neil, S.; Smallwood, R.H. The Epitheliome: modelling the social behaviour of cells. *BioSystems*, 2004b, 76, 89–100.

Walker, D.C.; Sun, T.; MacNeil, S.; Smallwood, R. Modeling the effect of exogenous calcium on keratinocyte and HaCat cell proliferation and differentiation using an agent-based computational paradigm. *Tissue Eng,* 2006a, 12(8), 2301–2309.

Walker, D.C.; Wood, S.; Southgate, J.; Holcombe, M.; Smallwood, R. An integrated agent-mathematical model of the effect of intercellular signaling via the epidermal growth factor receptor on cell proliferation. *J Theor Biol,* 2006b, 242, 774–789.

Weinan, E. and Engquist, B. Multiscale Modeling and Computation. *Notices of the AMS*, 2003, 50(9), 1062-1070.

Williams, K.A.; Saini, S.; Wick, T.M.. Computational fluid dynamics modelling of steady-state momentum and mass transport in a bioreactor for cartilage tissue engineering. *Biotechnology Progress*, 2002, 18, 951-963.

Wooldridge, J. and Jennings, NR. Intelligent Agents: Theory and Practice. *The Knowledge Engineering Review*, 1995, 10(2), 115-152.

Wooldridge, M. *An Introduction to MultiAgent Systems.* Chichester: John Wiley & Sons; 2002.

Wurm, M.; Lubei, V.; Caronna, M.; Hermann, M.; Margreiter, R.; Hengster, P. Development of a novel perfused rotary cell culture system. *Tissue Eng*, 2007, 13, 2761-2768.

Yu, X.; Botchwey, E.A.; Levine, E.M.; Pollack, S.R.; Laurencin, C.T. Bioreactor-based bone tissue engineering: the influence of dynamic flow on osteoblast phenotypic expression and matrix mineralization. *Proc. Natl. Acad. Sci. USA.* 2004; 101:11203.

Zeigler, B.; Kim, T.; Praehofer, H. *Theory of modeling and simulation: integrating discrete event and continuous complex dynamic systems*. San Diego: Academic Press; 2000.

Zimmermann, W.H.; Schneiderbanger, K.; Schubert, P.; Didie, M.; Munzel, F.; Heubach, J.F.; Kostin, S.; Neuhuber, WL; Eschenhagen, T. Tissue engineering of a differentiated cardiac muscle construct. *Circ Res*, 2002, 90, 223–230.

In: Computational Engineering
Editors: J. E. Browning and A. K. McMann

ISBN: 978-1-61122-806-9

Chapter 2

STOCHASTIC ANALYSIS AND NONLINEAR METAMODELING OF CRASH TEST SIMULATIONS AND THEIR APPLICATION IN AUTOMOTIVE DESIGN

I. Nikitin, L. Nikitina and T. Clees
Fraunhofer Institute for Algorithms and
Scientific Computing (SCAI),
Schloss Birlinghoven, Germany

ABSTRACT

Crash test simulations possess a stochastic component, related with physical and numerical instabilities of the underlying crash model and uncertainties of its control parameters.

Scatter analysis is a characterization of this component, usually in terms of scatter amplitude distributed on the model's geometry and evolving in time. This allows to identify parts of the model, possessing large scatter and most non-deterministic behavior. The purpose is to estimate reliability of numerical results and predict corresponding tolerances.

Causal analysis serves a determination of cause-effect relationships between events. In context of crash test analysis, this usually means identification of events or properties causing the scatter of the results. This allows to find sources of physical or numerical instabilities of the system and helps to reduce or completely eliminate them.

Sensitivity analysis considers dependence of numerical results on variations of control parameters. It allows to identify the parameters with the largest influence to the results and the parts (locations) of the model where the impact of such variations is considerable.

Metamodeling of simulation results allows to make prediction of the model behavior at new values of control parameters, for which a simulation has not been performed yet. These values of parameters are usually intermediate with respect to existing ones, for which simulations are available. Predictions are needed for analysis and optimization of product properties during the design stage and are usually completed with a control simulation performed at the optimal parameter values.

Statistical methods which allow for an efficient solution of these tasks are reviewed, and a novel method is introduced and its effiency demonstrated for benchmark cases.

Introduction

Today, simulation is an integral component of virtual product development. Simulation mainly comprises the solution of physically based models in the form of ordinary or partial differential equations. From the viewpoint of product development the real purpose is product optimization, and simulation is "only" one part of the process, though usually a quite costly one with respect to computational effort.

Optimization is a search for the best possible product with respect to multiple objectives (multi-objective optimization), e.g. total weight, fuel consumption and production costs, while simulation provides an evaluation of objectives for a particular sample of a virtual product. The optimization process usually requires a number of simulation runs; their results form a simulation dataset. To keep simulation time as short as possible, a "design of experiments" technique (DoE, [Tukey 1997]) is applied, where a space of design variables is sampled by a limited number of simulations. On the basis of these samples a mathematical model is constructed, e.g. a response surface [Donoho 2000], which describes the dependence between design variables and design objectives. Modern metamodels [Jones et al 1998; Keane et al 2005; Nikitina et al 2007-2008] describe not only the value of design objective but also its tolerance limits, which allow to control precision of the result. Moreover, not only single design objectives but also highly resolved simulation results from bulky datasets can be modeled [Nikitina et al 2008].

Imagine an interactive workspace where engineers can simulate in real time various designs of a motor hood and test its properties: Does a 0.5mm thin motor hood yet provide enough protection for passengers? How strong should the metal sheet be to fulfill new insurance requirements? How will the design of a car body and its crash behavior vary if a new composite material is used? How will the fuel consumption increase if the motor hood is strengthened? An interactive workspace would help engineers to vary different parameters easily and see their influences on a three-dimensional model of the vehicle in real time, allowing for an efficient evaluation of various designs and for finding the best solution with respect to the specific criteria applied.

However, today's industrial simulations have such a high complexity that they cannot be done in real time, in spite of the permanently increasing power of computers and common usage of parallel computations on clusters. Note that real-time performance means support of interactive work, i.e. a response time of a tenth of a second ideally. To approach this goal, innovative solution strategies in simulation, optimization and data analysis are required.

The other challenging problem is posed by inherently non-deterministic nature of physically based simulations used e.g. in crash test analysis. Typical indication is a large variation of simulation results under negligibly small variations of design parameters. Moreover, simulation results show sensitivity to such "internal" (numerical) details such as the distribution of computational processes on several CPUs and is generally not repeatable even for fixed values of parameters. Such stochastic behavior is related with physical and numerical instabilities which propagate through the model and amplify each other. Like a weather forecast possessing well known exponential instabilities, crash data analysis requires self-estimation of confidence limits to provide reliable predictions. In this way propagation of uncertainties in numerical simulations can be controlled.

Robust optimization accounts for uncertainties in the results to be on the safe side of prediction ("six-sigma design"). Since uncertainties of simulation results are distributed in space and time, it's natural to treat them as distributed random values, i.e. random fields [Khoshnevisan 2002]. This treatment poses an interesting problem of finding interdependencies, identifying correlated regions in bulky data, practically allowing to track causal relationships in the underlying physical model. Special methods [Thole, Mei 2003-2008; Thole 2010] have been developed for efficient detection of similarities and causal analysis of bulky crash data.

The above cited methods form a basis of the software tools DesParO [Nikitina et al 2008] and DiffCrash [Thole, Mei, 2003-2008] developed at Fraunhofer SCAI and have been subjects of international patent applications [Thole, Mei, 2003; Thole, Mierendorf 2009; Nikitina, Clees 2010].

Correlation Analysis of Bulky Data

Causal analysis is generally performed by means of statistical methods, particularly by estimation of correlation of events. It is commonly known that correlation does not imply causation (this logical error is often referred as "cum hoc ergo propter hoc": "with this, therefore because of this"). Instead, strong correlation of two events does mean that they belong to the same causal chain. Two strongly correlated events either have direct causal relation or they have a common cause, i.e. a third event in the past, triggering these two ones. This common cause will be revealed if the whole causal chain, i.e. a complete sequence of causally related events, will be reconstructed. Practical application of causal analysis requires formal methods for reconstruction of causal chains.

For this purpose, one can use the following standard statistical description:

Input: experimental data matrix x_{ij}, $i=1..N_{data}$, $j=1..N_{exp}$. Every column in this matrix forms one experiment, every row forms a data item varied in experiments.

Then every data item is transformed to a z-score vector [Larsen, Marx 2001]:

$$<x_i> = \Sigma_j\, x_{ij} / N_{exp}$$

with $dx_{ij} = x_{ij} - <x_i>$ and $|dx_i|=\mathrm{sqrt}(\Sigma_j\, dx_{ij}^2)$ and $z_{ij}=dx_{ij}/|dx_i|$

or by means of the equivalent alternative formula

$$z_{ij} = dx_{ij} / (\mathrm{rms}(x_i)\ \mathrm{sqrt}(N_{exp})),$$

with $\mathrm{rms}(x_i)=\mathrm{sqrt}(\Sigma_j\, dx_{ij}^2/\ N_{exp})$. Here $<x_i>$ is the mean of the i-th data item, dx_{ij} is the deviation from the mean of the i-th data item, and $\mathrm{rms}(x_i)$ is the root mean square deviation of the i-th data item.

In this way data items are transformed to N_{data} vectors in N_{exp} -dimensional space. All these z-vectors belong to an (N_{exp} -2)-dimensional unit-norm sphere, formed by intersection of a sphere $|z|=1$ with a hyperplane $\Sigma_j\, z_{ij}=0$.

An important role of this representation is the following. All z-vectors which are located close one to the other correspond to strongly correlated data items. Two such data items being displayed on 2D plot will show a linear dependence between the items. This plot is coincident up to a scaling with the plot of one z-vector versus the other, which is concentrated near $z_1=z_2$ due to approximate coincidence of z-vectors. Also in the case when two z-vectors are approximately opposite, they correspond to strongly correlated data items and linear data plot, approximately coincident with $z_1=-z_2$, up to a scaling.

The scalar product of two z-vectors is equal to Pearson's correlator since

$$(z_1,z_2) = \Sigma_j z_{1j} z_{2j} \text{ and } (z_1,z_2) = \text{corr}(x_1,x_2)$$

It is also related with the angular distance of the z-vectors on the sphere:

$$(z_1,z_2)=\cos \theta(z_1,z_2),$$

see Figure 6. Pearson's correlator, often used as a measure of correlation, indicates strong correlation in the limit of $\text{corr}\to 1$, corresponding to $z_1\to z_2$, $\theta\to 0$ or in the limit $\text{corr}\to -1$, corresponding to $z_1\to -z_2$, $\theta\to\pi$. The sign of Pearson's correlator shows the direction of dependence of data items: in the case corr=1, z_1 increases with increase of z_2, in the case corr=-1, z_1 decreases with increase of z_2. If not the direction of dependence is of interest, but only the fact of such dependence, one can formally consider a sphere of z-vectors as projective space, where opposite points are glued together and are considered as identical points. In this space approximately opposite vectors become approximately coincident.

The task of correlation clustering is to group all strongly correlated data items in one cluster and subdivide all data items to a few such clusters. This allows to identify the parts of data which behave similarly and to reduce the number of independent degrees of freedom of the system to few cluster centers. Clustering provides a compact representation of experimental data, suitable for direct analysis by user. Formally this representation can be defined as a reconstruction of original bulky data in terms of cluster centers which proceeds in two phases:

- phase I: determination of cluster centers,
- phase II: assignment of data items to cluster centers.

Correlation clustering becomes straightforward in z-space representation: one only needs to group together closely located z-vectors with any suitable general clustering technique. Many efficient clustering techniques are known to be applicable to this problem [Tan et al 2005]:

- density-based clustering
- k-means clustering
- dendrogram clustering
- spectral clustering

Details of these techniques will be presented below. Most of them possess as a control parameter a cluster diameter diam related to Pearson's correlator by the formula diam=arccos(corr).

Example of Bulky Data: Crash-Test Simulations

One of the largest challenges of crash-test data analysis is the huge amount of data items to be processed. Results of the commonly used finite-element-based crash-test simulation are presented in groups of different information levels:

- Points in space (nodes) storing x,y,z-coordinates, velocities, accelerations, stress tensor components, material properties, etc. The number of nodes is very large (millions).
- Elements (solids, shells, beams etc) combine nodes, can also store material properties and other values. The number of elements is also very large (millions).
- Parts (b-pillar, tunnel, etc) combine elements, the number of parts is moderate, e.g. several thousands.
- Timesteps combine parts, the number of timesteps is moderate, e.g. several hundreds.

The total number of data items is then

$$N_{data} = (N_{nodes} * N_{values\_per\_node} + N_{elements} * N_{values\_per\_element}) * N_{timesteps}$$

is usually hundreds of millions. Special methods are required to support processing of such a huge amount of data.

Number of experiments N_{exp} is usually varied from dozens to hundreds. In order to keep computational cost reasonable, N_{exp} is required to be as minimal as possible. The following observation allows keeping N_{exp} moderate. In the case of strong correlation (e.g. $|corr|>0.7$), the fact of interdependence of data items is directly visible on raw data plots, e.g. 2-dimensional plots displaying one data item versus the other one. Detection of such interdependences has the first priority, since they are present in the data themselves and do not rely upon models and assumptions made in the analysis. The analysis serves only a detection of such interdependencies. Although this detection is still a sophisticated task due to the large amount of data items (up to 10^8) and an even larger number of pairs of these data items (up to 10^{16}), only a few experiments are needed to detect strong correlation. The statistical error of correlator in the limit $|corr| \to 1$ can be estimated as

$$rms(corr) = (1-|corr|)\ sqrt(6 / N_{exp}).$$

This estimation shows that $N_{exp} = 25$ experiments are sufficient to achieve a precision of rms=0.1 at $|corr|=0.8$, while about $N_{exp} = 300$ experiments are necessary for the same precision of rms=0.1 for the smaller value of $|corr|=0.3$. The value $N_{exp} = 25$ also corresponds to an acceptable computational effort and is therefore optimal for analysis.

Clustering Techniques

The *density-based clustering* technique estimates the density of data items in z-space by introducing a discretization of z-space to a finite set of hypercubes. The hypercubes should cover the unit-norm sphere where all data items are located. The hypercube size should correspond to the optimal cluster diameter diam=arccos(corr). In every hypercube, matching z-score vectors are counted. Although the total number of hypercubes is huge due to the large dimension of z-space – this phenomenon is referred as "curse of dimensionality" [Donoho 2000] –, only a small portion of hypercubes is filled. Practically, the number of filled hypercubes is much less than the number of data items. For efficient access to so structured data, one can use known methods of multidimensional indexing, based on binary search trees. Such methods support rapid search, insert and erase operations, with a typical operation count of $O(\log(N_{elements}))$. An implementation of the methods can be found in standard software libraries such as C++ standard template library STL. After processing all data items and storing hypercube count data in a binary tree, the following procedure is applied, see Figure 1. An hypercube A with a maximal count is selected, and its center is defined as the cluster center. Then all hypercubes B are removed possessing a correlation with A stronger than a specified weak limit (e.g. those with |corr|>0.5). The procedure is repeated for the remaining hypercubes C until the maximal count becomes sufficiently small. Density-based clustering requires $O(N_{cubes})$ memory and between $O(N_{data} * N_{exp})$ and $O(N_{data} * \log(N_{cubes}))$ floating point operations, where in practice $N_{cubes} << N_{data}$.

The *k-means clustering* technique, also known as Voronoj iteration, is illustrated by Figure 2. Its initialization requires specifying the expected number of clusters k and randomly selecting k cluster centers. Then each data item is assigned to the nearest cluster center, and new cluster centers are recomputed. This step is repeated iteratively until the cluster assignments do not change any more. This technique allows to obtain a minimal average cluster diameter, which therefore becomes a function of number of clusters k: <diam>(k). The optimal value of k corresponds to <diam>(k*)=arccos(corr). This can be achieved by initially selecting a small value of k (underestimation, e.g. k=5) and sequentially incrementing k until optimal condition is met. Otherwise one can select a large value for k (overestimation, e.g. k=1000) and decrement k to the optimal value. k-means has moderate requirements w.r.t. memory $O(N_{data} * N_{exp})$, and computational complexity $O(N_{data} * N_{exp} * k * N_{iter})$, depending on the number of iterations.

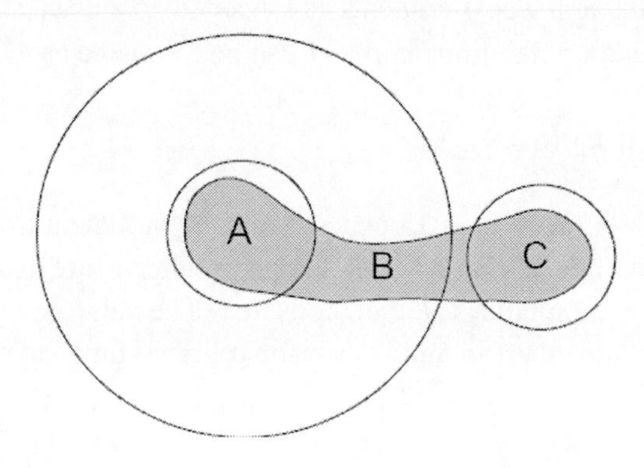

Figure 1. Density-based clustering.

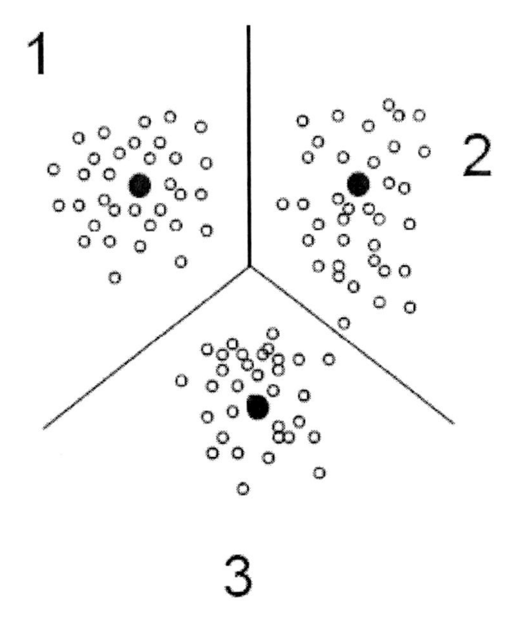

Figure 2. k-means clustering.

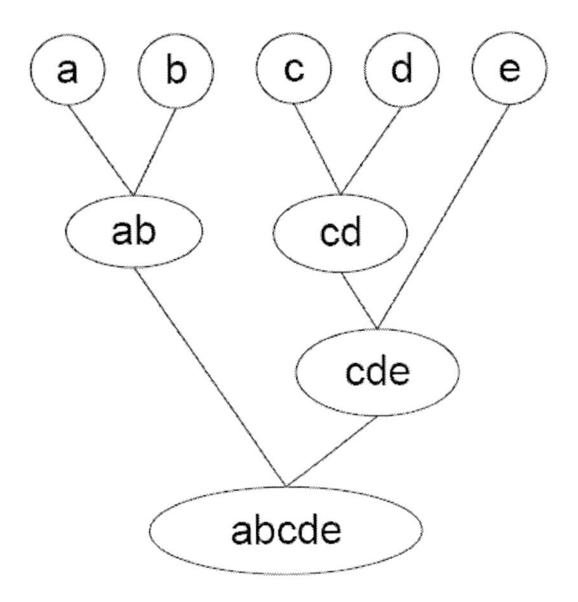

Figure 3. Dendrogram clustering.

The *dendrogram clustering* technique, also known as agglomerative hierarchical clustering, stores a distance matrix for clusters d(X,Y). For the case of correlation clustering it can be related with correlation matrix, defined in terms of scalar products of corresponding z-vectors, e.g.

$$d(x,y) = \arccos |\mathrm{corr}(x,y)| = \arccos |(z_x,z_y)|.$$

So-defined distance between data items is then used to determine distance between clusters, by means of one of these alternative definitions:

- single linkage: $d(X,Y) = \min d(x,y)$, $x \in X$, $y \in Y$,
- complete linkage: $d(X,Y) = \max d(x,y)$, $x \in X$, $y \in Y$,
- average linkage: $d(X,Y) = \text{avg}\, d(x,y)$, $x \in X$, $y \in Y$.

The method defines original points as initial clusters, then applies to them the following procedure. A pair of clusters is found possessing minimal distance. This pair is merged to a new cluster, and the distance matrix is updated accordingly (merging two rows corresponding to clusters X and Y to a new row using the linkage formula above, complementing merge of X and Y columns from matrix symmetry). The procedure is repeated until all data will be merged into one cluster, or a stopping criterion will be met. For example, if the minimal distance exceeds the critical diameter diam=arccos(corr), iterations can be stopped. A sequence of merged clusters defined by this procedure can be graphically displayed as a tree-like diagram, called dendrogram, see Figure 3. A stopping criterion cuts off a tree not reaching the root, leaving a number of clusters with user-controlled diameter. Dendrogram clustering is memory- and computationally expensive, since it requires to store a $O(N_{data}{}^2)$ distance matrix and to perform up to $O(N_{data}{}^3)$ operations for processing. For bulky data, it can be applied only in a combination with a special subsampling procedure, as described below. On the other hand, this technique is non-iterative and has a predictable completion time.

Spectral clustering is based on a special matrix decomposition, known as singular value decomposition (SVD). The Z-matrix is a row-wise collection of z-vectors for all data items and is decomposed as follows:

$$Z=U\Lambda V^T$$

where Λ is a diagonal matrix of size $N_{exp} * N_{exp}$, U is a column-orthogonal matrix of size $N_{data} * N_{exp}$, V an orthogonal square matrix of size $N_{exp} * N_{exp}$:

$$U^TU=1,\ V^TV=VV^T=1.$$

A computationally efficient method to perform this decomposition in our practical case with $N_{data} >> N_{exp}$ is to find the Gram matrix:

$$G=Z^TZ\ ,$$

perform its spectral decomposition:

$$G=V\Lambda^2V^T\ ,$$

and compute the remaining U-matrix with the formula:

$$U=ZV\Lambda^{-1}.$$

The Gram matrix has a moderate size of $N_{exp} * N_{exp}$ which allows to load the whole G-matrix to memory and apply any suitable method for spectral matrix decomposition (e.g. Jacobi method). Practically, the computational complexity of this step is negligible compared with the other steps. Computation of G-matrix consists of sequential reading of data items and incrementing the elements of G-matrix, e.g. for n-th data item ($n=1.. N_{data}$):

$$G_{ij} = G_{ij} + z_{ni}\, z_{nj},\ i,j=1..\ N_{exp},$$

requiring $O(N_{data} * N_{exp}^{\ 2})$ floating point operations and $O(N_{exp}^{\ 2})$ amount of memory. Computation of U-matrix requires to read the database once more, performing $O(N_{data} * N_{exp} * N_{clust})$ floating point operations and storing $O(N_{exp}^{\ 2})$ data items. Further optimizations are possible, based on symmetry of G-matrix, storage of only $N_{exp} * N_{clust}$ piece of V-matrix etc. Here, N_{clust} is the number of selected spectral clusters, practically $N_{clust} << N_{exp}$, see Figure 4.

Selection of clusters can be performed using Parseval's criterion. The SVD spectrum of the Z-matrix is represented by elements of the diagonal matrix Λ. These values are non-negative and sorted in non-ascending order. If all these values starting from a certain number $N_{clust}+1$ are omitted (i.e. set to zero), the resulting reconstruction of Z-matrix will have a deviation dZ. L_2-norm of this deviation gives

$$err^2 = \Sigma_{ij}\, dz_{ij}^{\ 2} = \Sigma_{k>Nclust}\, \Lambda_k^{\ 2}$$

i.e. the sum of squares of omitted spectral values. This formula allows controlling precision of reconstructed Z-matrix. Usually Λ_k rapidly decreases with k, and a few first Λ values give sufficient precision.

In this way, spectral clustering provides an efficient representation of the Z-matrix as a linear combination of a small number of U-vectors. It is convenient to scale these vectors by entries of the diagonal Λ-matrix, writing $\Psi=U\Lambda$. Column vectors of Ψ are called spectral clusters or modes, while column vectors of the V-matrix define corresponding cluster centers in z-space.

In a separable case, when all data items can be assigned to N_{clust} exactly uncorrelated clusters, the cluster centers in z-space, found by any other clustering method, become mutually orthogonal (since scalar product is equal to correlator) and identical with columns of V-matrix. U-vectors in this case will have non-overlapping content, taking non-zero values on data items from corresponding cluster and zero values outside.

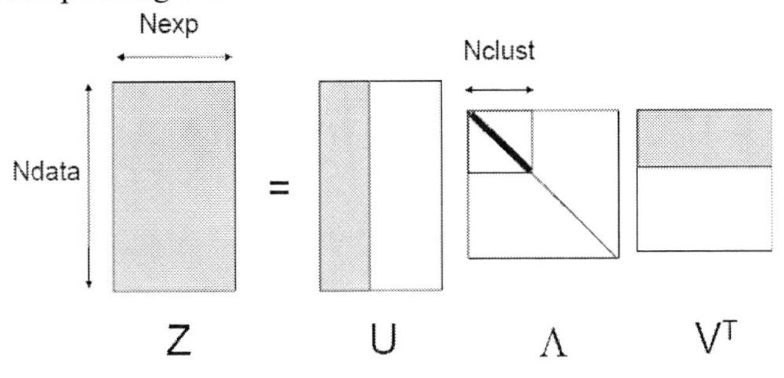

Figure 4. Spectral clustering.

Generally, there can be data items weakly correlated with several cluster centers. In this case, U-vectors will overlap, reflecting the fact that data items belong simultaneously to several clusters. Also, cluster centers can be weakly correlated with one another and can be approximately, but not exactly orthogonal. In this case, V-vectors will represent an orthogonal basis in z-space nearest to cluster centers, in this way approximating positions of actual cluster centers. The reconstruction of the Z-matrix provided by spectral clustering remains exact also in this general case and can be used e.g. for interpolation of bulky data (see below).

Spectral Clustering and Other Spectral Decompositions

The formulae above implement spectral clustering for the Z-matrix. This is mathematically equivalent to a spectral decomposition of the correlation matrix

$$C=ZZ^T=U\Lambda^2U^T$$

here, the matrix C has size $N_{data} * N_{data}$ and is composed of scalar products of z-score vectors for every pair of data items (equal to correlator of data items). Due to a huge size of this matrix, it is practically impossible to compute, store or spectrally decompose it directly. However, the above method provides its correct and computationally efficient spectral decomposition. One can use in the above formulae instead of the Z-matrix the dX-matrix, composed of dx-vectors (see above). The resulting decomposition will be equivalent to spectral decomposition of the covariance matrix

$$C_2= (dX)\,(dX)^T,$$

which has the same size $N_{data} * N_{data}$ and is composed of average values $cov_{ij}=\langle dx_i\, dx_j\rangle$ (with a normalization factor N_{exp}). The matrices C and C_2 differ only in scaling factors, equal to scatter rms(x_i), applied to every row and column. Generally, one can assign an arbitrary weight per data item, e.g. for the purpose of selecting more interesting regions of data and compute SVD for matrix QZ, where Q is a diagonal matrix containing these weights. This is equivalent to the spectral decomposition of the matrix

$$C_q= QZZ^TQ.$$

The results of spectral clustering for different selections of weights are equivalent up to orthogonal transformation, i.e. after a variation of weights, $\Xi=Q^{-1}U\Lambda$ is replaced by $\Xi\to\Xi\Omega$, with Ω being a $N_{exp} * N_{exp}$ orthogonal matrix:

$$\Omega\Omega^T=\Omega^T\Omega=1.$$

In the separable case described above, the matrices C, C_2, C_q have block-diagonal structure and their decompositions become independent on weights (Ω=1).

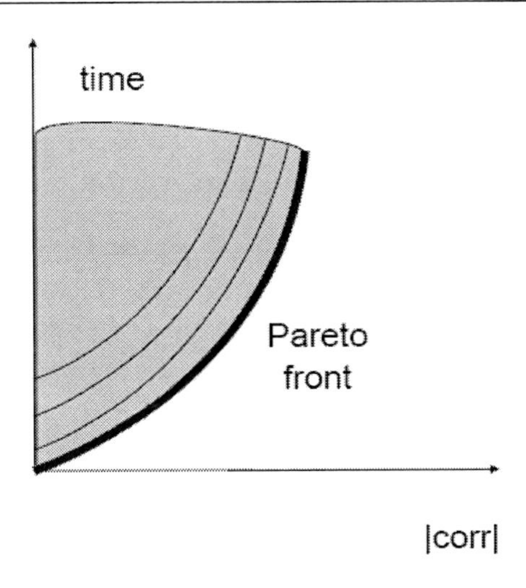

Figure 5. Pareto ranking.

Chaining phenomenon: a problem which can occur in application of clustering techniques is appearance of chains of data items where consequential items are strongly correlated but beginning and end items are not correlated. Such chains should not be merged in one cluster. Density-based clustering includes a special algorithm for elimination of strongly correlated chain members. The k-means technique attempts to minimize the average cluster diameter and will tend to break the chains and assign them to different cluster centers. The dendrogram technique is free from the chaining phenomenon if the complete linkage scheme is used. Spectral clustering produces cluster centers which are orthogonal and uncorrelated by definition.

Specific of projective space: it has been already noted that only the absolute value of the correlator is important, and opposite z-vectors can be identified by means of gluing the z-sphere to (N_{exp} -2)-dimensional projective space. The density-based technique can be modified to count opposite z-vectors in the same hypercube applying a transformation $z' = \{z, z_1 \geq 0; -z, z_1 < 0\}$ and covering only half of the sphere by hypercubes. k-means can be similarly modified by redefinition of the assignment criterion: a z-vector is assigned to cluster center z_c if the absolute value $|(z,z_c)|$ is maximal, and a z-vector is reverted $z \rightarrow -z$ if $(z,z_c)<0$. The dendrogram technique takes the structure of the projective space automatically into account by a special definition of the non-Euclidean distance, so that opposite z-vectors produce a zero entry in the distance matrix. Spectral clustering merges opposite cluster centers in one cluster, assigning them to one V-vector.

Acceleration of Clustering Methods

Spectral clustering allows to process data items sequentially one after the other and has the minimal memory requirements. On the other hand, it requires considerable computational effort for filling the Gram matrix.

Density based clustering also allows to process data items sequentially, since it stores only density information per hypercube. Its problem, however, is a rapidly increasing number of hypercubes and proportional memory requirements. Practically there are only a few hypercubes with millions of data items, and millions of hypercubes contain only a few data items. Only the most populated hypercubes are interesting for the analysis. The following intermediate "cleaning procedure" can reduce the memory usage. During the filling of hypercubes, periodically, e.g. after every million data items and N_1=100 times totally, hypercubes which have too few data items, e.g. less than N_2=100, are removed. As a result, only a few really populated hypercubes survive the procedure. The amount of data per hypercube can be underestimated by N_1*N_2=10000 which is much less than the amount of data items in the hypercube (a million). This procedure, similar to the percentage hurdle voting rule, allows a significant reduction of the total memory usage of the algorithm.

Other clustering methods require processing simultaneously all data items and, for optimal performance, need to load in memory the whole data matrix or even the larger distance matrix. Due to their enormous sizes this is practically impossible. The following observation allows applying these methods nevertheless. Since the most interesting clusters contain millions of data items, their centers and the actual density of data in them can be reliably determined on much smaller statistics than the full set. Subsampling of data items is a selection of random subset in the full set of data items, which can be used for determination of cluster centers. The selection should be really random and not correlated with node, part or state structure of the dataset. After determining the cluster centers, the full set of data items is processed once to assign them to the clusters. This approach is similar to a public opinion poll by selecting a representative sample of the population. It allows reducing the memory usage and accelerating computation significantly (by a factor of thousands).

Points of Interest

Correlation clustering delivers cluster centers in form of z-vectors. In addition to cluster centers, the user can define other points of interest involving her/his engineers knowledge of the problem. For example, certain control points on the car body can be important from safety requirements and should be included into the POI list.

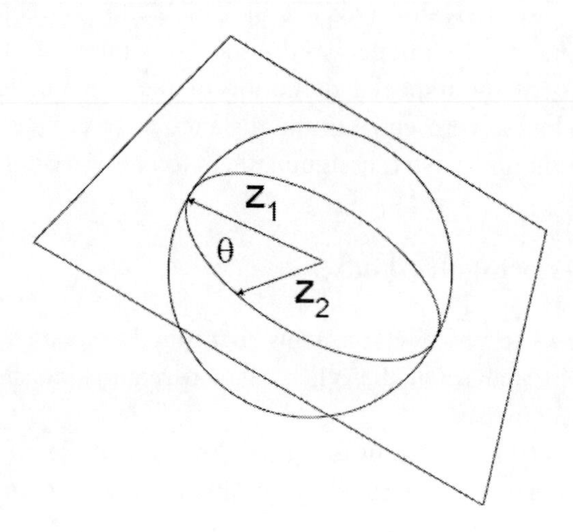

Figure 6. z-score vectors.

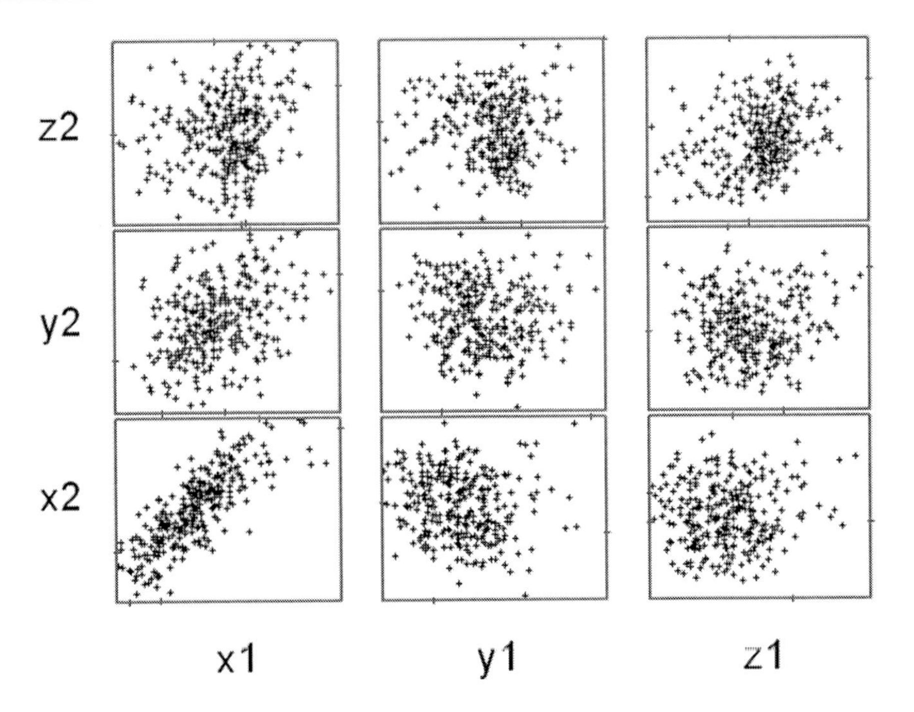

Figure 7. Raw data plots for all component combinations.

After the POI list has been set up, the main pass of the computation is applied. It computes correlator corr(POI,*) of all items from the POI list with all data items, as well as scatter amplitude rms(*) for all data items. This step involves one-time reading of all data and is normally one of the most time consuming parts of the analysis.

The following data organization is necessary for the main pass. The data items are stored in uncompressed or compressed form with a random access to blocks of data (e.g. states). Experiments can be stored in a few separate files (e.g. N_{exp} =25), which should be opened simultaneously. This structure supports an optimal order of data reading, which allows rapid sampling of one data item, i.e. x_{ij}, at i fixed, j varied from 1 to N_{exp}. This vector is transformed to the z-space, where a few of computations are performed to find rms, corr and other necessary statistical measures of the data item. The resulting measures are written blockwise to a file, if necessary, in compressed form. For example, the data can be written in the data format of the original experiments, ready for visualization with any suitable post-processor program.

To achieve a more compact storage and comprehensive visualization, for tensor structures, such as 3-vectors and 3x3 matrices, an effective norm can be selected and written to a file as a scalar value. For example, scatter of vector values has three components, and its Euclidean norm appear to be most convenient:

$$s_i = \mathrm{sqrt}(\Sigma_{k=1,2,3}\ \mathrm{rms}(x_{3i+k})^2),$$

For the correlator, the maximal absolute value appears more convenient

$$\mathrm{mcorr}_{ij} = \max_{k1,k2=1,2,3} |\mathrm{corr}_{3i+k1,3j+k2}|,$$

since $|mcorr_{ij}| \to 1$ implies that at least one of the tensor components represents strong correlation. In the formulae above, data items are ordered by triples: $(X_1,Y_1,Z_1,X_2,Y_2,Z_2,...)$, see Figure 9. On this figure highest hierarchical levels are timesteps t1,t2,..., each of them contains nodes, represented by triples of coordinates.

The lost component information can be recovered afterwards by observing raw data plots for any pair of selected nodes. For example, for a destination point from the POI list and for a corresponding source point found in causal analysis, raw data can be extracted from the data set and displayed as 3x3 plots including all component combinations, see Figure 7. Random access to data items allows to accelerate such data extraction. It accelerates also clustering algorithms, utilizing a subsampling procedure, since only a small selected subset of data must be read. Also, extraction of user-defined POI for computation of correlator is accelerated. Cluster centers which are formally added to the POI list are already presented as z-score vectors and do not need to be extracted. Other parts of computation require only sequential data access.

Summarizing, the effort for processing each of the clustering techniques mentioned includes, in addition to purely computational effort, also input/output operations. For determination of cluster centers one needs $O(N_{subsamp} * N_{exp})$ reading operations, if subsampling is applied, else $O(N_{data} * N_{exp})$ reading operations without subsampling. For output of clustering data one needs $O(N_{data} * N_{exp})$ operations for one-pass reading of the whole database and $O(N_{data} * N_{clust})$ operations to write the clustering results. All input/output operations can be accelerated using data compression algorithms.

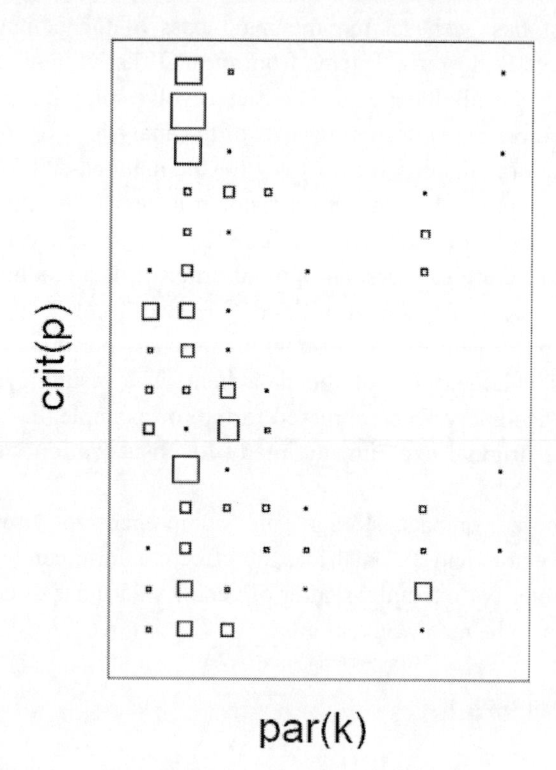

Figure 8. Sensitivity matrix. Sizes of squares represent values of matrix elements, color can be used to display sign of the dependence.

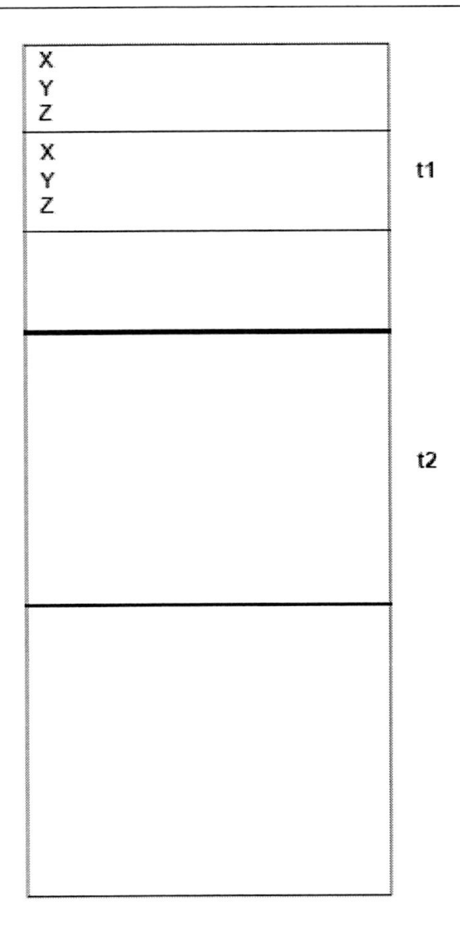

Figure 9. An example of data ordering in experimental data matrix.

Partition Diagram

A partition diagram is a subdivision of a geometrical model to regions of influence: each data item from the geometrical model is checked against user-specified and/or cluster POI lists to find closest POI in the sense of correlation distance dist=arccos(|corr|). A single scalar value is stored per state and node, the integer part of which is a number of the closest POI and the fractional part of which is related with the correlator to this POI:

$$val = numPOI + (1-|corr|)$$

For example, val=3.2 means that a given point has strongest correlation with the third POI at the level corr=0.8, while val=3.7 means that strongest correlation is still achieved for the third POI, although the correlation level is weak, corr=0.3.

Representation of data in a form of partition diagram shows - at a glance – the correlation structure of the whole data set. Selecting an appropriate visualization range in available postprocessing tools such as GNS Animator, one can investigate further details e.g. search a time step where the region of influence for a particular POI originates, in order to identify the source of scatter related with this POI.

Scatter and Causal Analysis

A practical problem of causal analysis in crash-test simulations is often not a removal of a primary cause of scatter, which is usually the crash event itself. It is more an observation of propagation paths of the scatter, with a purpose to prevent this propagation, by finding regions where scatter is amplified (e.g. break of a welding point, pillar buckling, slipping of two contact surfaces etc). Therefore it is necessary to check an amplitude of scatter rms(x) along the causal chain. Also, small cause can have large effect and formally earliest events in the causal chain can have microscopic amplitude ("butterfly effect"). Again, it is reasonable to search for amplifying factors and try to eliminate them, not the microscopic sources.

Causal chains can be defined as regions in space-time possessing a large correlator (e.g. $|corr|>0.7$) with a given destination point. Such regions can be directly visualized using computed correlation data. A selection of the best representative path inside the chain can be done using 2-dimensional multi-objective optimization for finding parts with maximal correlation at earliest possible state. This gives a curve (Pareto front) on a 2-dimensional plot (time,|corr|) representing the causal chain, see Figure 5. One can also add to the representative set a neighbourhood of the Pareto front, which can be determined with Pareto ranking methods.

Linear Sensitivity Analysis

Sensitivity analysis considers dependence of the numerical result on variation of control parameters par_k, k=1...Npar. In crash problems such control parameters can be e.g. thicknesses of car body components, initial velocity of collision, material properties etc. The purpose of linear sensitivity analysis is, based on the determination of a Jacobi matrix $J_{ik}=\partial x_i/\partial par_k$ or, in a different normalization, a sensitivity matrix $S_{ik}=J_{ik} rms(par_k)$, a decision which parameters mostly influence the data. Our correlation data allow solving rapidly also this task. Particularly, in case of a uniform design of experiment (DoE) with a random and uncorrelated change of parameters, we have

$$S_{ik} = rms(x_i)\ corr(x_i, par_k)$$

i.e. the sensitivity matrix is proportional to the correlator of data item with variation of parameter. To compute these data, one only needs to include data items representing the variation of parameters to the list of POI, in addition to cluster centers and user-defined POIs. However, in case of many parameters, the resulting sensitivity matrix is too bulky for direct representation to the user. It is better to compute the sensitivity matrix only for cluster centers, since they represent main tendencies in data:

$$S_{pk} = rms(clust_p)\ corr(clust_p, par_k)$$

This gives a comprehensible matrix of size N_{clust} x N_{par}, the entries of which can be directly visualized to detect most influencing parameters, e.g. in the form shown in Figure 8.

A similar formula can be obtained also for a non-uniform DoE and a correlated change of parameters:

$S_{pk} = \mathrm{rms}(clust_p)\ \Sigma_{m=1..Npar}\ \mathrm{corr}(clust_p,par_m)\ C^{-1}_{mk}$,

where C_{mk}=corr(par_m,par_k) is the inter-correlation of parameters.

Non-Linear Sensitivity Analysis and Metamodeling of Bulky Data

Metamodeling is an interpolation or approximation method applied to existing simulations to predict the result of a new simulation performed at intermediate values of parameters. Often non-linear methods are used to catch non-linear parametric dependencies. The usage of radial basis functions (RBFs) is quite convenient, i.e. a representation of the form

$$f(par)= \Sigma_{i=1..\ Nexp}\ c_i\ \Phi(|par-par_i|)$$

is used, where par_i are coordinates of the i-th run in the space of design variables; f is the objective value, interpolated to a new point par; Φ() are special functions, depending only on the Euclidean distance between the points par and par_i. The coefficients c_i can be obtained by solving a linear system

$$f_i = \Sigma_j\ c_j\ \Phi(|par_i\ -par_j|)\ ,$$

where f_i are the user-provided data, which could be single criteria or bulky simulation results, associated with i-th simulation run. The solution can be found by direct inversion of the moderately sized N_{exp} * N_{exp} system matrix $\Phi_{ij} = \Phi\ (|par_i\ -par_j|)$. The result can be written in a form of weighted sum:

$$f(par) = \Sigma_i\ w_i(par)\ f_i,$$

with the weights $w_i(par) = \Sigma_j\ \Phi^{-1}_{ij}\ \Phi\ (|par-par_j|)$. This representation makes an interpolation of bulky simulation data possible at a rate (computational complexity) linear with respect to the size of the dataset.

A suitable choice for the RBF is the multi-quadric function [Buhmann 2003]

$$\Phi\ (r)= \mathrm{sqrt}(c^2+r^2),$$

where c is a constant defining smoothness of the function near data point par=par_i. In between data points, the result of the interpolation is a cumulative effect of many contributions and, except of this smoothness, there is hardly any influence of coefficient c to the result. Therefore adjustment procedures (learning) is not necessary, in contrast to other interpolation techniques like Kriging and neural networks [Keane et al 2005]. RBF interpolation can also be combined with polynomial detrending, adding a polynomial part P:

$$f(par)= \Sigma_{i=1..Nexp}\ c_i\ \Phi\ (|par-par_i|) + P(par).$$

This allows reconstructing exactly polynomial (including linear) dependencies and generally improving precision of interpolation. Formulae for coefficients c_i and interpolation weights w_i can be updated accordingly.

Metamodeling is employed to complete sensitivity analysis for non-linear dependencies, where instead of the S_{pk} matrix other quantities can be used as sensitivity measures. For example

$$S^{max}{}_{pk} = \max_{par\,k} |S_{pk}|(par_k),$$

with the maximum taken over the k-th central cross-section in parameter space:

$$par_j=(par_j^{min}+par_j^{max})/2,\ j\neq k,\ par_k\in[par_k^{min},par_k^{max}].$$

Also the sign of S_{pk} can be tested along the cross section, and the global sign of the dependence can be introduced, equal to +1 if all S_{pk} are positive on cross-section, -1 if all S_{pk} are negative and 0 for sign changing S_{pk}. This information can be easily presented graphically, e.g. in the form shown in Figure 8.

Possible interaction terms can be estimated similarly using higher order derivatives.

Metamodeling is also used for the solution of optimization problems. In practical applications, engineers are confronted with simultaneous optimization of multiple objectives, coming from different types of simulations (so called multidisciplinary optimization problems). It is often desirable to interpolate the whole model (even on highly resolved geometries), instead of some scalar optimization criteria only, in order to have a possibility to inspect the (Pareto-)optimal design in full detail. Such kind of bulky data metamodeling can be significantly accelerated using correlation clustering described in this chapter.

This possibility is based on the fact that, for bulky crash data, the (approximate) rank of the data matrix is usually much smaller than the number of experiments, rank<< N_{exp}. The rank of a matrix is the maximal number of linearly independent columns and also rows in it. This fact means that not the whole z-space is used, but a small subspace in it. To identify this subspace, any of the above described clustering methods can be used. The basis of this subspace can be spanned on cluster centers. Because all clustering techniques except of spectral clustering can produce a non-orthogonal basis, the decomposition formula should generally use a dual basis:

$$z_{ij}=\Sigma_{k=1..Nexp,p=1..Nclust}\,(z_{ik}\,zc_{pk})\,zc'_{pj},\ zc'_{pj}=\Sigma_{q=1..Nclust}\,G^{-1}{}_{pq}\,zc_{qj},$$

$$G_{pq}=\Sigma_{n=1..Nexp}\,(zc_{pn}\,zc_{qn}).$$

Here, zc' is the dual basis to zc, G is the Gram matrix of the basis zc. Rewriting this formula in terms of correlators, and expressing dx

$$dx_{ij} = \mathrm{sqrt}(N_{exp})\mathrm{rms}(x_i)\ \Sigma_{p=1..Nclust}\ \mathrm{corr}(clust_p,x_i)zc'_{pj}$$

as a result, the whole data matrix can be expressed as a sum of N_{clust} terms found earlier with the corr(clustPOI,*) algorithm. The advantage of such representation is that N_{clust} =rank<<

N_{exp} leading to a reduction of storage and the possibility to accelerate all processing operations applied to this matrix. Particularly, non-linear metamodeling of bulky data represents an interpolated result as a weighted sum:

$$dx^*_i = \Sigma_{j=1..Nexp}\, dx_{ij}\, w_j.$$

with the clustering representation above this gives

$$dx^*_i = \mathrm{sqrt}(N_{exp})\mathrm{rms}(x_i)\, \Sigma_{p=1..Nclust}\, \mathrm{corr}(clust_p,x_i)\, g_p,\; g_p=\Sigma_{j=1..Nexp}\,(zc'_{pj}\, w_j)$$

providing an acceleration factor of N_{exp} / N_{clust} , compared with the direct computation of the weighted sum.

A similar representation can be written for spectral clustering:

$$dx^*_i = \mathrm{sqrt}(N_{exp})\mathrm{rms}(x_i)\, \Sigma_{p=1..Nclust}\, \Psi_{ip}\, g_p\, ,\; g_p=\Sigma_{j=1..Nexp}\,(V_{jp}\, w_j)$$

$$\Psi=ZV,\; \Psi_{ip} = (z_i,zc_p) = \mathrm{corr}(clust_p,x_i),$$

where columns of V define centers of clusters and

$V_{jp} = zc_{pj} = zc'_{pj}$, because the V-basis is self-dual.

A further advantage of RBF metamodeling is that it allows not only to interpolate data in a new parameter point, but also to estimate precision of this interpolation. In an existing data point, this precision can be estimated using a known cross-validation procedure: the data point is removed, data are interpolated to this point and compared with the actually measured value at this point. Formally, for this comparison one needs to construct a new metamodel where the data point is removed. For an RBF metamodel with detrending this step can be performed analytically, resulting in the following direct formula:

$$err_i = f_{interpol}(par_i) - f_{actual}(par_i) = -c_i/(\Phi^{-1})_{ii}$$

where c_i and Φ were defined above. In between data points, the so-estimated error can be interpolated, using e.g. the same RBF weights w_i. The error formula above is applicable also for bulky data, since

$$c_{pi} = \Sigma_{j=1..Nexp}\,(\Phi^{-1})_{ij}\, dx_{pj},\; p=1..\, N_{data},\; i=1..\, N_{exp}$$

is represented as a weighted sum of data items and correlation clustering allows accelerating its computation as well.

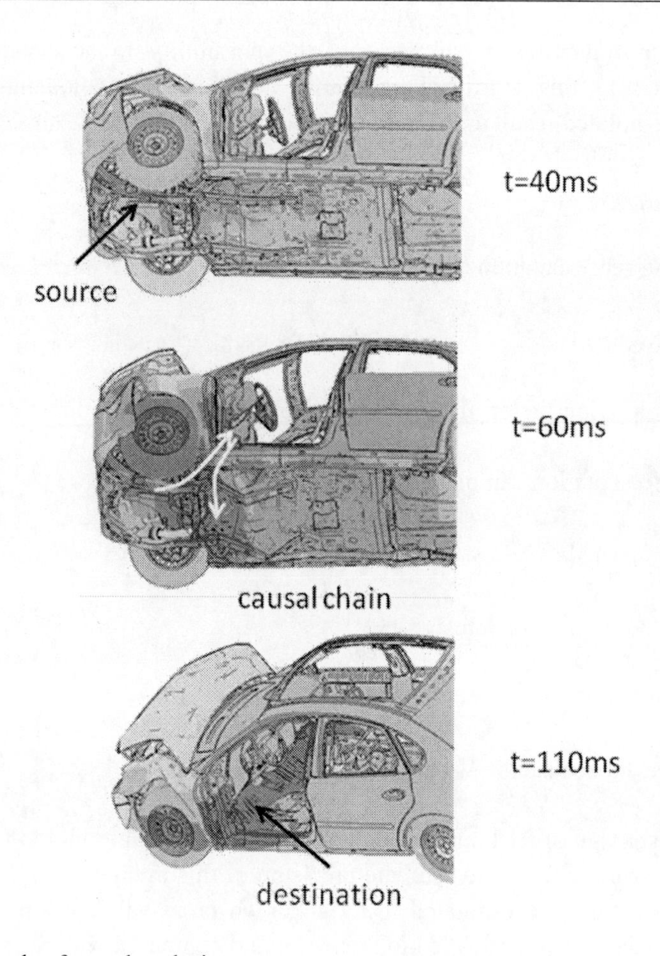

Figure 10. An example of causal analysis.

Real-Life Examples

Example 1: Ford Taurus Crash Test

The considered crash model (Figure 10,11) contains 1 million nodes and 32 timesteps; 25 simulations have been performed.

Crash intrusions in foot rooms of the driver and passenger are commonly considered as critical safety characteristics of car design. These characteristics possess numerical uncertainties, the analysis of which falls in the subject of this chapter. For a user-specified destination point, selected in driver's foot room, depicted in Figure 10 (bottom), correlators have been computed using the method described in this chapter. Color (grey code) displays the value of the correlator. A source of the scatter, showing an increased value of the correlator, has been identified at an earlier time step. The source is related with a buckling of the longitudinal structure depicted in Figure 10 (top). Large correlation values propagate along the causal chain depicted in Figure 10 (in the middle). This chain, detected with the Pareto-front algorithm, shows propagation of the scatter from the source towards the steering wheel, and further to the destination point in the foot room.

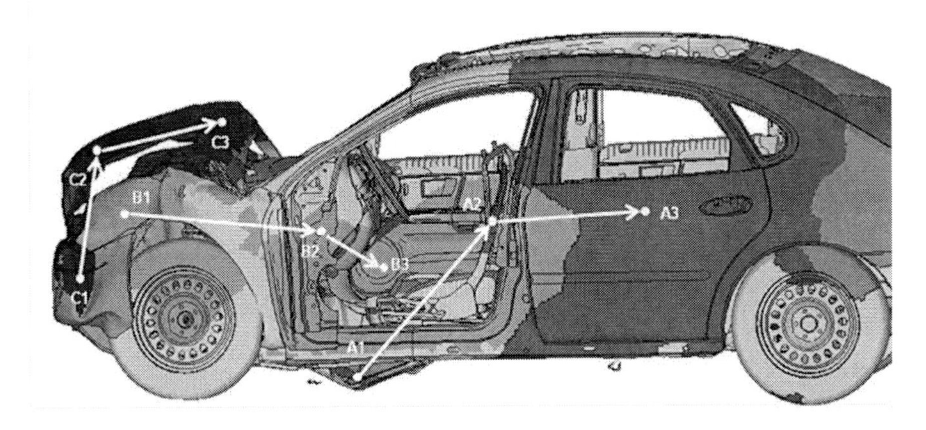

Figure 11. An example of partition diagram.

The partition diagram, depicted in Figure 11, shows at a glance the distribution of correlators in the model. Several correlation clusters have been detected, in particular:

- Cluster A1-A2-A3 corresponds to a buckling of the bottom longitudinal structure, the influence of which propagates through the driver's seat and extends to the whole left-rear side of the car.
- Cluster B1-B2-B3 starts on left wheel house and propagates through the steering column to the right side of the car.
- Cluster C1-C2-C3 starts on the bumper at the moment of crash with the barrier and extends over the motor hood.

Color (grey code) is used to distinguish clusters on the model.

Correlation clustering algorithms have been recently integrated into the software tool DiffCrash.

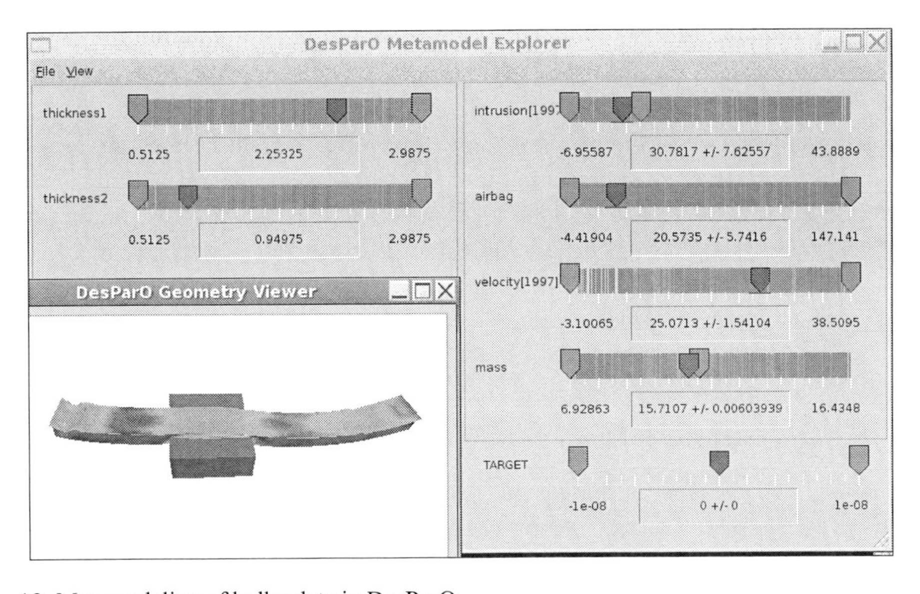

Figure 12. Metamodeling of bulky data in DesParO.

Example 2: Volkswagen Lupo Crash Test and NVH Analysis

The sensitivity matrix shown in Figure 8 results from the analysis reported in [Nikitina et al 2007]. In that paper, non-linear sensitivity analysis based on RBF metamodeling is applied for solving a multidisciplinary constrained optimization problem at Volkswagen AG. The problem includes 12 criteria evaluated in 3 types of simulation experiments: front crash of the car with a rigid wall, crash with a deformable barrier (both using the PamCrash software) and NVH analysis (using the Nastran simulation software). Optimization objectives are: maximal intrusions in different critical points, maximal acceleration, first eigenfrequencies and torsional stiffness. The main objective is to reduce total mass of the car body; other criteria are considered as constraints: their values should not be worse than for initial design. Design variables are 10 thicknesses of 15 parts of the car body (certain parts share the same thickness value). Initially, 270 simulation designs have been generated using a Latin Hypercube DoE with ±10% boundaries from the initial design. The pattern of interdependencies between parameters and criteria shown in the matrix in Figure 8 allows to identify the most influencing parameters for each type of simulations. For example, this matrix shows that almost all criteria are very sensitive to the variation of the second parameter, e.g. the fifth criterion depends only on this parameter, while the ninth criterion depends also on the first parameter. Further analysis, involving global sign of dependence, shows that the ninth criterion increases with an increase of the second parameter and decreases with an increase of the first one. This information has become very helpful for the optimization process, which has allowed to reduce the total mass of the vehicle by 1.5kg keeping all constraints satisfied.

Example 3: Audi B-pillar Crash Test

The model shown in Figure 12 contains 10 thousand nodes and 45 timesteps; 101 simulations have been performed. Two parameters are varied representing thicknesses of two layers composing the B-pillar part. The purpose is to find a Pareto-optimal combination of parameters simultaneously minimizing the total mass of the part and crash intrusion in a selected middle point. While the methods of multi-objective optimization combined with RBF metamodeling are applied to solve this problem, spectral clustering is used for a compact representation of the bulky simulation results. Spectral clustering has recently been integrated into the interactive optimization tool DesParO [Nikitina et al 2008] and allows to accelerate algorithms responsible for interpolation of bulky data by two orders of magnitude to fulfill real-time requirements. In DesParO, the user can interactively change parameter values in order to immediately see their impact on optimization criteria and - facilitated by means of spectral clustering – even on complete simulation results.

CONCLUSION

In this chapter, we have discussed a general strategy for the analysis of crash-test data which allows to accelerate scatter analysis, causal analysis, sensitivity analysis, and metamodeling of bulky data. It consists of the following steps:

- a number of crash-test simulations or real experiments is performed and stored in the database;
- correlation clustering of data items in the database is performed;
- a list of points of interest (POI) is formed including cluster centers, user defined points and variations of design parameters;
- one-pass computation of scatter and correlator of all data items with the POI list is performed;
- a partition diagram is set up;
- scatter and causal analysis is performed based on Pareto ranking of scatter and correlator data;
- linear sensitivity analysis is performed using correlator of parameter variations with the POI list;
- non-linear sensitivity analysis and metamodeling of bulky data are performed based on an reconstruction of the database in terms of correlation clusters.

Applications of this strategy has been demonstrated for real-life examples from the field of automotive design.

The strategy can more generally be used for applications from engineering or natural sciences with geometrical background. Besides studies of parameter sensitivity, examples include

- forming (e.g. for analysis of spring-back or clicker phenomena)
- casting and molding (e.g. for analysis of blow holes arising in a solidification process)
- computational fluid dynamics (e.g. for backtracking eddies or finding and backtracking sources of numerical instabilities, e.g. due to cracks, tiny in- or outflows dominating the overall flow field or due to inadequate local discretization),
- medicine (e.g. for backtracking instable results inside soft matter, e.g. organs, due to irradiation processes such as focused ultra-sound)
- semiconductor process simulation (e.g. for backtracking of mechanical stress peaks to areas with non-matching crystal lattices or for the analysis of doping evolution after deposition or injection due to chemical reactions and diffusion),
- semiconductor device simulation (e.g. for finding and backtracking current leaks),
- network simulation and circuit simulation (if combined with geometrical information; e.g. for backtracking leaks),
- electromagnetic fields, electromagnetic compatibility (e.g. for backtracking local sensitivities due to wave evolution or the analysis of numerical instabilities).

ACKNOWLEDGMENTS

This work was supported by the FhG Internal Programs under Grants No. MAVO 816 450 (CAROD) and MAVO 817759 (HIESPANA). We are grateful to Clemens-August Thole (SCAI) for helpful discussions, to Daniela Steffes-lai (SCAI) for data preparation, to Roel

Kersten, Juergen Bruns, Michael Taeschner and Georg Dietrich Eichmueller (VW) for fruitful collaboration.

REFERENCES

Buhmann N.D., *Radial basis functions: theory and implementations*, Cambridge University Press, 2003.

Donoho D. L., High-Dimensional Data Analysis: The Curses and Blessings of Dimensionality. Lecture on August 8, 2000, to the American Mathematical Society "Math Challenges of the 21st Century". Available from http://www-stat.stanford.edu/~donoho/.

Jones D.R., Schonlau M., Welch W.J., Efficient Global Optimization of Expensive Black-Box Functions, *Journal of Global Optimization*, vol.13, 1998, pp.455-492.

Keane A. J., Leary S.J., Sobester A., On the Design of Optimization Strategies Based on Global Response Surface Approximation Models, *Journal of Global Optimization*, vol.33, 2005, pp. 31-59.

Khoshnevisan D., *Multiparameter Processes - An Introduction to Random Fields*, Springer 2002.

Larsen R.J., Marx M.L., *An Introduction to Mathematical Statistics and Its Applications*, Prentice Hall 2001.

Nikitina L., Clees T., *Apparatus and method for analyzing bulky data*, 2010, patent pending.

Nikitina L., Nikitin I., Steffes-lai D., Thole C.-A., Kersten R., Bruns J., Constrained optimization with DesParO, in Proc. Conf. Virtual Product Development in Automotive Engineering, Prien, 21-22 March 2007.

Nikitina L., Nikitin I., Stork A., Thole C.-A., Klimenko S., Astakhov Y., Towards Interactive Simulation in Automotive Design, *Visual Computer* 2008, V24 pp.947-953.

Tan P.-N., Steinbach M., Kumar V., *Introduction to Data Mining,* Addison-Wesley, 2005.

Thole C.-A., Mei L., Reason for scatter in simulation results. In *Proceedings of the 4th European LS-Dyna User Conference*. Dynamore, Germany, 2003.

Thole C.-A., Mei L., Data analysis for parallel car-crash simulation results and model optimization. *Simulation modelling practice and theory,* 16(3), pp.329–337, 2008.

Thole C.-A., Advanced Mode Analysis for Crash Simulation Results, in *Proc. 11th International LS-DYNA Users Conference*, Dearborn, USA, June 06-08, 2010.

Thole C.-A., Mei L., *Procedure for analysing crash testing data dispersion*, 2003, patent No. EP1338883.

Thole C.-A., Mierendorff H., DPMA No.102009057295.3, 2009, patent pending.

Tukey J.W., *Exploratory Data Analysis*, Addison-Wesley, London, 1997.

In: Computational Engineering
Editors: J. E. Browning and A. K. McMann
ISBN: 978-1-61122-806-9

Chapter 3

Brains Meet Topological Quantum Computers: Quantum Neural Computation

***Vladimir G. Ivancevic*[1,*] *and Tijana T. Ivancevic*[2,†]**
[1]Defence Science & Technology Organisation, Australia
[2]Society for Nonlinear Dynamics in Human Factors, Australia

Abstract

Classical computing systems perform classical computations (i.e., Boolean operations, such as AND, OR, NOT gates) using devices that can be described classically (e.g., MOSFETs). On the other hand, quantum computing systems perform classical computations using quantum devices (quantum dots), that is, devices that can be described only using quantum mechanics. Any information transfer between such computing systems involves a state measurement. This review paper describes this information transfer at the edge of topological chaos, where mysterious quantum-mechanical linearity meets even more mysterious brain's nonlinear topological complexity, in order to perform a super–high–speed and error–free computations.

Keywords: neurodynamics, topological quantum computation, topological chaos

1. Introduction

In biological neural networks, signals are transmitted between neurons by electrical pulses (action potentials or spike trains) travelling along the axon. These pulses impinge on the afferent neuron at terminals called synapses. These are found principally on a set of branching processes emerging from the cell body (soma) known as dendrites. Each pulse occurring at a synapse initiates the release of a small amount of chemical substance or neurotransmitter which travels across the synaptic cleft and which is then received at postsynaptic receptor sites on the dendritic side of the synapse. The neurotransmitter becomes bound to molecular sites here which, in turn, initiates a change in the dendritic membrane potential. This

*E-mail address: Vladimir.Ivancevic@dsto.defence.gov.au
†E-mail address: Tijana.Ivancevic@alumni.adelaide.edu.au

postsynaptic potential (PSP) change may serve to increase (hyperpolarize) or decrease (depolarize) the polarization of the postsynaptic membrane. In the former case, the PSP tends to inhibit generation of pulses in the afferent neuron, while in the latter, it tends to excite the generation of pulses. The size and type of PSP produced will depend on factors such as the geometry of the synapse and the type of neurotransmitter. Each PSP will travel along its dendrite and spread over the soma, eventually reaching the base of the axon (axonhillock). The afferent neuron sums or integrates the effects of thousands of such PSPs over its dendritic tree and over time. If the integrated potential at the axonhillock exceeds a threshold, the cell fires and generates an action potential or spike which starts to travel along its axon. This then initiates the whole sequence of events again in neurons contained in the efferent pathway.

On the other hand, *artificial neural networks* (ANNs) are computational systems loosely modeled on the human brain (see, e.g. [5]). The field goes by many names, such as *connectionism*, *parallel distributed processing*, *neuro–computing*, natural intelligent systems, *machine learning* algorithms and ANNs. It is an attempt to simulate within specialized hardware or sophisticated software, the multiple layers of simple processing elements called neurons. Each neuron is linked to certain of its neighbors with varying coefficients of connectivity that represent the strengths of these connections. Learning is accomplished by adjusting these strengths to cause the overall network to output appropriate results.

In this paper we will show the possibilities of merging (recurrent) ANNs with modern topological quantum computers.

2. Classical Neurodynamics

Although currently there is a large variety of models for ANNs, they all share eight major aspects: (i) A set of processing units, or 'neurons', represented by a set of integers; (ii) An activation for each unit, represented by a vector of time-dependent functions;
(iii) An output function for each unit, represented by a vector of functions on the activations; (iv) A pattern of connectivity among units, represented by a matrix of real numbers indicating connection strength; (v) A propagation rule spreading the activations via the connections, represented by a function on the output of the units; (vi) An activation rule for combining inputs to a unit to determine its new activation, represented by a function on the current activation and propagation; (vii) A learning rule, which can be either unsupervised such as Hebbian [3], or supervised, such as *backpropagation* [2, 5], for modifying connections based on experience, represented by a change in the 'synaptic weights' based on any number of variables; (viii) An environment which provides the system with experience, represented by sets of activation vectors for some subset of the units.

To give a brief introduction to classical neurodynamics, we start from the fully recurrent, $N-$dimensional, RC transient circuit, given by a nonlinear vector differential equation [2, 4, 5]:

$$C_j \dot{v}_j = I_j - \frac{v_j}{R_j} + w_{ij} f_i(v_i), \qquad (i, j = 1, ..., N), \tag{1}$$

where $v_j = v_j(t)$ represent the activation potentials in the jth neuron, C_j and R_j denote input capacitances and leakage resistances, synaptic weights w_{ij} represent conductances,

I_j represent the total currents flowing toward the input nodes, and the functions f_i are sigmoidal.

Geometrically, equation (1) defines a smooth autonomous vector–field $X(t)$ in ND neurodynamical phase–space manifold M, and its (numerical) solution for the given initial potentials $v_j(0)$ defines the autonomous neurodynamical phase–flow $\Phi(t) : v_j(0) \to v_j(t)$ on M.

In AI parlance, equation (1) represents a generalization of three well–known recurrent NN models (see [2, 4, 5]):
(i) Continuous Hopfield model [7],
(ii) Grossberg ART–family cognitive system [10, 11], and
(iii) Hecht–Nielsen counter–propagation network [8, 9].

Physiologically, equation (1) is based on the Nobel–awarded *Hodgkin–Huxley equation* of the neural action potential (for the single squid giant axon membrane) as a function of the conductances g of sodium, potassium and leakage [12, 13]:

$$C\dot{v} = I(t) - g_{Na}(v - v_{Na}) - g_K(v - v_K) - g_L(v - v_L),$$

where bracket terms represent the electromotive forces acting on the ions.

The *continuous Hopfield circuit* model [7]:

$$C_j\dot{v}_j = I_j - \frac{v_j}{R_j} + T_{ij}u_i, \qquad (i, j = 1, ..., N), \tag{2}$$

where u_i are output functions from processing elements, and T_{ij} is the inverse of the resistors connection–matrix becomes equation (1) if we put $T_{ij} = w_{ij}$ and $u_i = f_i[v_j(t)]$.

The Grossberg *analogous ART2 system* is governed by activation equation:

$$\varepsilon\dot{v}_j = -Av_j + (1 - Bv_j)I_j^+ - (C + Dv_j)I_j^-, \qquad (j = 1, ..., N),$$

where A, B, C, D are positive constants (A is dimensionally conductance), $0 \leq \varepsilon << 1$ is the fast–variable factor (dimensionally capacitance), and I_j^+, I_j^- are excitatory and inhibitory inputs to the jth processing unit, respectively.

General *Cohen–Grossberg activation equations* [10] have the form:

$$\dot{v}_j = -a_j(v_j)[b_j(v_j) - f_k(v_k)m_{jk}], \qquad (j = 1, ..., N), \tag{3}$$

and the *Cohen–Grossberg theorem* ensures the global stability of the system (3). If

$$a_j = 1/C_j, \qquad b_j = v_j/R_j - I_j, \qquad f_j(v_j) = u_j,$$

and constant $m_{ij} = m_{ji} = T_{ij}$, the system (3) reduces to the Hopfield circuit model (2).

The Hecht–Nielsen *counter–propagation network* is governed by the activation equation [8, 9]:

$$\dot{v}_j = -Av_j + (B - v_j)I_j - v_j\sum_{k\neq j} I_k, \qquad (j = 1, ..., N),$$

where A, B are positive constants and I_j are input values for each processing unit.

Provided some simple conditions are satisfied, namely, say symmetry of weights $w_{ij} = w_{ij}$, non–negativity of activations v_j and monotonicity of transfer functions f_j, the system

(1) is globally asymptotically stable (in the sense of Liapunov energy functions). The fixed–points (stable states) of the system correspond to the fundamental memories to be stored, so it works as content–addressable memory (AM). The initial state of the system (1) lies inside the basin of attraction of its fixed–points, so that its initial state is related to appropriate memory vector. Various variations on this basic model are reported in the literature [2, 4], and more general form of the vector–field can be given, preserving the above stability conditions.

3. Topological Quantum Computers

The concept of *quantum computation*, was first stated by Feynman [14] and Benioff [15], and formalized by Deutsch [16], Bernstein and Vazirani [17], and Yao [18]. For a thorough recent overview, see [1]. Briefly, *quantum computer* is a computation device that makes direct use of distinctively quantum–mechanical phenomena, such as *superposition* and *entanglement*,[1] to perform operations on data. Whilst in a conventional computer information is stored as bits, in a quantum computer it is stored as quantum binary digits, or *qubits*. The basic principle of quantum computation is that the quantum properties can be used to represent and structure data, and that quantum mechanisms can be devised and built to perform operations with these data [19].

Unfortunately, quantum computers seem to be extremely difficult to build. The qubits are typically expressed as certain quantum properties of *trapped particles*, such as individual atomic ions or electrons. But their superposition states are exceedingly fragile and can be spoiled by the tiniest stray interactions with the ambient environment, which includes all the material making up the computer itself. If qubits are not carefully isolated from their surroundings, such disturbances will introduce errors into the computation. Most schemes to design a quantum computer therefore focus on finding ways to minimize the interactions of the qubits with the environment. Researchers know that if the error rate can be reduced to around one error in every 10,000 steps, then *error–correction procedures* can be implemented to compensate compensate for decay of individual qubits. Constructing a functional machine that has a large number of qubits isolated well enough to have such a low error rate is a daunting task that physicists are far from achieving [21, 1].

For this reason, a few researchers are pursuing a very different, topological way to build a quantum computer. In their approach the delicate quantum states depend on *topological*

[1] *Quantum entanglement*, a phenomenon referred to by E. Schrödinger as 'the essence of quantum physics', is a property of quantum superpositions involving more than one system. Just as two classical bits can be in any of four states $(00, 01, 10, 11)$, the general quantum state of two qubits is a superposition of the form $c_{00}|00\rangle + c_{01}|01\rangle + c_{10}|10\rangle + c_{11}|11\rangle$ and the quantum state of N qubits can be represented by a complex–valued vector with $2N$ components. This is the basis of the exponential superiority of quantum computation: instead of N Boolean registers, one has $2N$ complex variables, even though there are only N physical switches. But to be computationally useful, the joint quantum state must be 'non-separable'. A separable state can be expressed as an abstract product of individual states:

$$|00\rangle = |0\rangle_A|0\rangle_B, \qquad |00\rangle + |01\rangle = |0\rangle_A(|0\rangle + |1\rangle)_B.$$

However, the so–called *Bell state*, $|00\rangle + |11\rangle$, cannot be factorized in this way, and is therefore non-separable. The entanglement of a state is a measure of its non-separability, and arguably represents the fundamental resource used in quantum computation [20].

properties of a quantum system.[2] The so–called *topological quantum computer* is a theoretical quantum computer that employs 2D quasi-particles called *anyons*, whose *world lines* cross over one another to form *braids* (see Figure 1) in a $(1+2)-$space-time.

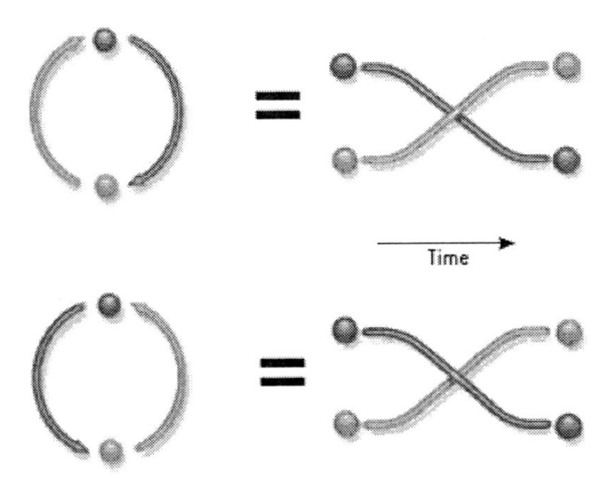

Figure 1. Braiding: two anyons can encounter a clockwise swap (top) and a counterclockwise swap (bottom). These two moves in a plane generate all the possible braidings of the world lines of a pair of anyons (modified and adapted from [21]).

These braids form the *logic gates* that make up the quantum computer. The advantage of a quantum computer based on quantum braids over using *trapped quantum particles* is that the former is much more stable. While the smallest perturbations can cause a quantum particle to decohere and introduce errors in the computation, such small perturbations do not change the topological properties of the quantum braids (see Figure 2). This is like the effort required to cut a string and reattach the ends to form a different braid, as opposed to a ball (representing an ordinary quantum particle in 4D space-time) simply bumping into a wall. While the elements of a topological quantum computer originate in a purely mathematical realm, recent experiments indicate these elements can be created in the real world using semiconductors made of gallium arsenide near absolute zero and subjected to strong magnetic fields.

Anyons are quasi-particles in a 2D space. Anyons are not strictly fermions or bosons, but do share the characteristic of fermions in that they cannot occupy the same state. Thus, the world lines of two anyons cannot cross or merge. This allows braids to be made that make up a particular circuit. In the real world, anyons form from the excitations in an electron gas in a very strong magnetic field, and carry fractional units of magnetic flux in a particle–like manner. This phenomenon is called the *fractional quantum Hall effect*.[3] The

[2]Recall that topology is called a rubber–sheet geometry, i.e., a geometrical study of properties that are unchanged when an object is smoothly deformed, by actions such as stretching, squashing and bending but not by cutting or joining. It embraces such subjects as *knot theory*, in which small perturbations do not change a topological property. For example, a closed loop of string with a knot tied in it is topologically different from a closed loop with no knot. The only way to change the closed loop into a closed loop plus knot is to cut the string, tie the knot and then reseal the ends of the string together. Similarly, the only way to convert a topological qubit to a different state is to subject it to some such violence.

[3]The fractional quantum Hall effect (FQHE) is a physical phenomenon in which a certain system behaves as if it were composed of particles with charge smaller than the elementary charge. Its discovery and explanation

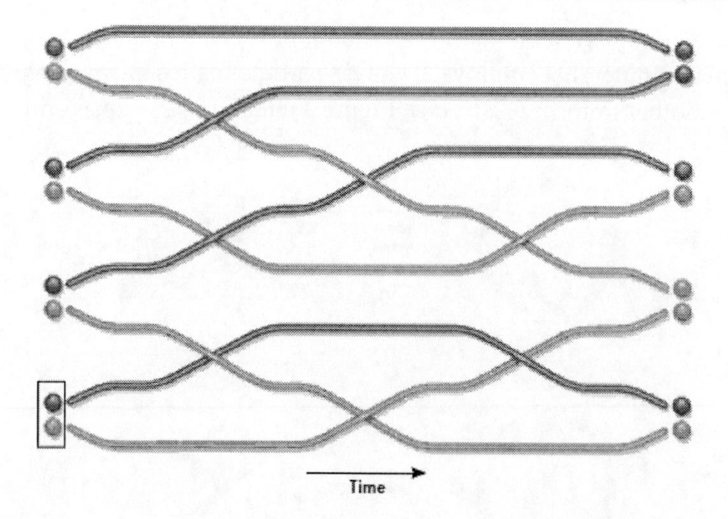

Figure 2. Computing with braids of anyons. First, pairs of anyons are created and lined up in a row to represent the qubits, or quantum bits, of the computation. The anyons are moved around by swapping the positions of adjacent anyons in a particular sequence. These moves correspond to operations performed on the qubits. Finally, pairs of adjacent anyons are brought together and measured to produce the output of the computation. The output depends on the topology of the particular braiding produced by those manipulations. Small disturbances of the anyons do not change that topology, which makes the computation impervious to normal sources of errors (modified and adapted from [21]).

electron 'gas' is sandwiched between two flat plates of gallium arsenide, which create the 2D space required for anyons, and is cooled and subjected to intense transverse magnetic fields.

When anyons are braided, the transformation of the quantum state of the system depends only on the topological class of the anyons' trajectories (which are classified according to the *braid group*, see Figure 3. Therefore, the quantum information which is stored in the state of the system is impervious to small errors in the trajectories.

The first proposal for topological quantum computation is due to Alexei Kitaev in 1997. In 2005, Fields Medalist Michael Freedman and collaborators proposed a quantum Hall device which would realize a *topological qubit*. The problem of finding specific braids for doing specific computations was tackled in 2005 by N.E. Bonesteel and collaborators, who showed explicitly how to construct a so–called controlled NOT (or CNOT) gate to an accuracy of two parts in 10^3 by braiding six anyons (see Figure 4). A CNOT gate takes

were recognized by the 1998 Nobel Prize in Physics. The FQHE is a manifestation of simple collective behavior in a 2D system of strongly–interacting electrons. At particular magnetic fields, the electron gas condenses into a remarkable state with liquid–like properties. This state is very delicate, requiring high quality material with a low carrier concentration, and extremely low temperatures. As in the integer quantum Hall effect, a series of plateaus forms in the Hall resistance. Each particular value of the magnetic field corresponds to a filling factor (the ratio of electrons to magnetic flux quanta) $\nu = p/q$, where p and q are integers with no common factors. In particular, *fractionally charged quasi-particles* are neither bosons nor fermions and exhibit *anyonic statistics*. The FQHE continues to be influential in theories about topological order. Certain fractional quantum Hall phases appear to have the right properties for building a topological quantum computer.

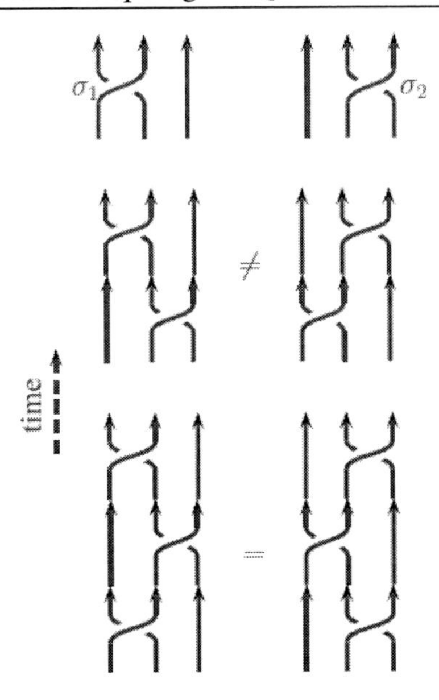

Figure 3. Graphical representation of elements of the braid group. Top: the two elementary braid operations σ_1 and σ_2 on three anyons. Middle: non-commutativity is shown here as $\sigma_2\sigma_1 \neq \sigma_1\sigma_2$; hence the braid group is non-Abelian. Bottom: the *braid relation*: $\sigma_i\sigma_{i+1}\sigma_i = \sigma_{i+1}\sigma_i\sigma_{i+1}$ (modified and adapted from [22]).

two input qubits and produces two output qubits. Those qubits are represented by triplets (green and blue) of so–called *Fibonacci anyons*. The particular style of braiding, leaving one triplet in place and moving two anyons of the other triplet around its anyons, simplified the calculations involved in designing the gate [1].

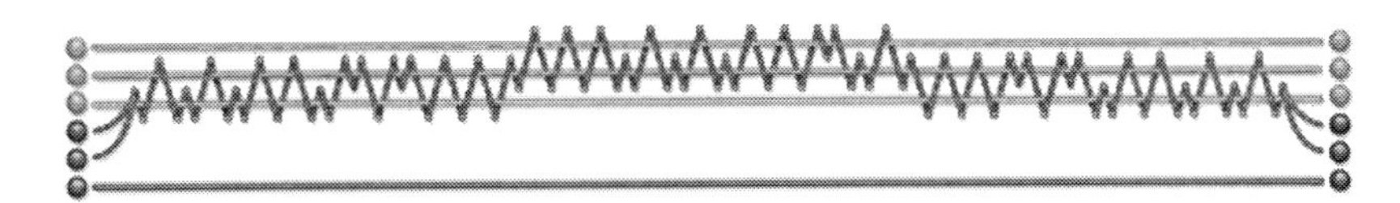

Figure 4. Recently, a *quantum logic gate* known as a *CNOT–gate* has been produced by a complicated braiding of six anyons. This braiding produces a CNOT gate that is accurate to about 10^3 (modified and adapted from [21]).

Topological quantum computers are equivalent in computational power to other standard models of quantum computation, in particular to the *quantum circuit* model and to the *quantum Turing machine* model. That is, any of these models can efficiently simulate any of the others. Nonetheless, certain algorithms may be a more natural fit to the topological quantum computer model. For example, algorithms for evaluating the *Jones polynomial* were first developed in the topological model, and only later converted and extended in the discrete quantum circuit model.

Even though quantum braids are inherently more stable than trapped quantum particles, there is still a need to control for error inducing thermal fluctuations, which produce random stray pairs of anyons which interfere with adjoining braids. Controlling these errors is simply a matter of separating the anyons to a distance where the rate of interfering strays drops to near zero. It has been estimated that the error rate for a CNOT operation of a qubit state could be as low as 10^{-30} or less. Although this number has been criticized as being strongly overstated, there is nonetheless good reason to believe that topologically protected systems will be particularly immune to many sources of error that plague other schemes for quantum information processing. Simulating the dynamics of a topological quantum computer may be a promising method of implementing fault–tolerant quantum computation even with a standard quantum information processing scheme.

To build a topological quantum computer requires one additional complication: the anyons must be what is called non-Abelian, or, non-commutative, which means that the order in which particles are swapped is important.[4] If *non-Abelian anyons* actually exist,[5] topological quantum computers could well leapfrog discrete quantum computer designs in the race to scale up from individual qubits and logic gates to fully fledged machines more deserving of the name 'computer'. Carrying out calculations with quantum knots and braids, a scheme that began as an esoteric alternative, could become the standard way to implement practical, error-free quantum computation [21].

4. Computation at the Edge of Chaos and Multi-Agents Dynamics

Depending on the connectivity, the so–called *recurrent neural networks*, consisting of simple computational units, can show very different types of dynamics, ranging from totally ordered (linear–like behavior) to chaotic. Using the the *mean–field theory* approach with *evolving Hamming distance*, Bertschinger and Natschläger analyzed how the type of dynamics (ordered or chaotic) exhibited by randomly connected networks of threshold gates driven by a time-varying input signal depended on the parameters describing the distribution of the connectivity matrix [37]. In particular, the authors calculated the critical boundary in *parameter space* where the transition from ordered to chaotic dynamics takes place. Employing a recently developed framework for analyzing real-time computations, they showed that only near the critical boundary could such networks perform complex computations on time series. This result strongly supports conjectures that dynamical systems that are capable of doing complex computational tasks should operate near the *edge of chaos*, that is, the transition from ordered to chaotic dynamics (see, e.g. [23, 25]).

In particular, the authors pointed out the following: (i) Dynamics near the critical line are a general property of input–driven dynamical systems that support complex real-time computations; and (ii) Such systems can be used by training a simple readout function to approximate any arbitrary complex filter. This suggests that to exhibit sophisticated information processing capabilities, an adaptive system should stay close to *critical dynamics*. Since the dynamics is also influenced by the statistics of the input signal, these results

[4] This is similar to the rotation group (in which the order of rotations matters) versus the translation group (in which the order of translations does not matter).

[5] Non-Abelian anyons probably exist in certain gapped 2D systems, including *fractional quantum Hall effect* and possibly also ruthenates, *topological insulators*, rapidly rotating *Bose–Einstein condensates*, *quantum loop gases*/string nets.

indicate a new role for plasticity rules: stabilize the dynamics at the critical line. Such rules would then implement what could be called *dynamical homeostasis*, which could be achieved in an unsupervised learning manner.

4.1. Topological Braiding Chaos in Multi-Agents Dynamics

To introduce topological braids mixing into quantum computation, we will firstly describe intuitively how braids arise in multi-agent dynamics. Figure 5-(a) shows the orbits of four particle-like agents in a circular 2D domain. Figure 5-(b) shows the *world-line* of the same orbits: They are plotted in a 3D graph, with time flowing vertically upward. The diagram in Figure 5-(b) depicts a physical braid made up of four strands. No strand can go through another strand as a consequence of the deterministic motion of the agents[6]. Moreover, the mathematical definition of a braid requires that strands cannot loop back: Here this simply means that the agents cannot travel back in time. We will say that two braids are equivalent if they can be deformed into each other with no strand crossing other strands or boundaries [26]. We are mostly interested in characterizing the level of *entanglement* of agents trajectories.

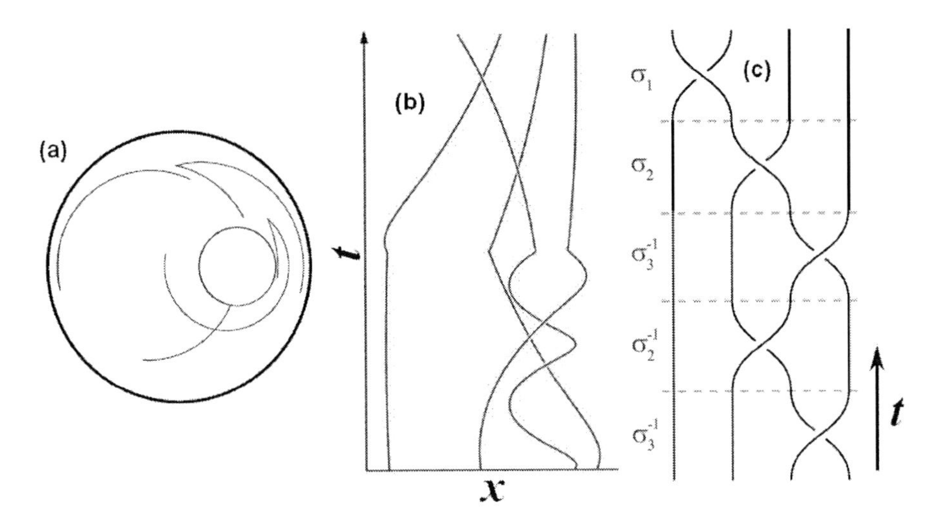

Figure 5. Arising of braids in multi-agent dynamics: (a) A zoom-into the orbits of four agents in a circular 2D domain. (b) The same orbits, lifted to a space-time diagram in 3D, with time flowing from bottom to top. (c) The standard braid diagram corresponding to (b) (modified and adapted from [26].

Since we can move the strands, is convenient to draw braids in a normalized form, as shown in Fig. 5-(c) for the braid in Fig. 5-(b). Such a picture is called a *braid diagram*. The important thing is that we record when crossings occur, and which agent was behind and which was in front. It matters little how we define 'behind,' as long as we are consistent. In Fig. 5-(c) the horizontal dashed lines also suggest that we can divide the braid into a sequence of elementary crossings, known as *generators*.

[6]they never occupy the same point at the same time

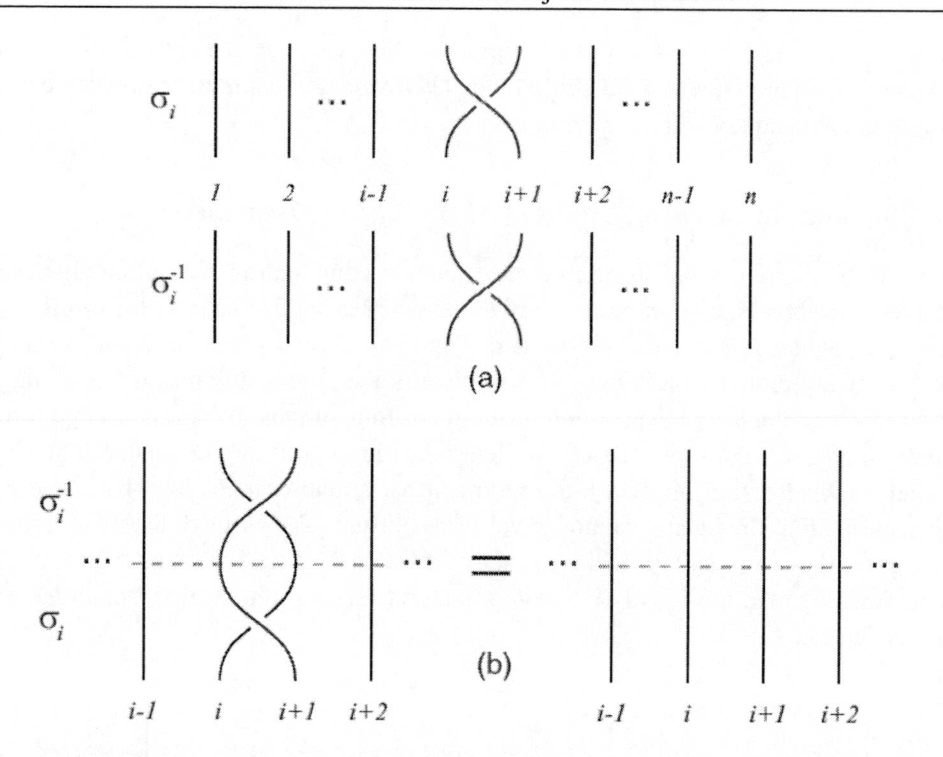

Figure 6. (a) Braid group generator σ_i, corresponding to the clockwise interchange of the string at the ith and $(i + 1)$th position, counted from left to right. Its inverse σ_i^{-1} involves their counterclockwise exchange. (b) The concatenation of σ_i and σ_i^{-1} gives the identity braid (modified and adapted from [26].

Figure 6-(a) shows the definition of σ_i, which denotes the clockwise interchange of the ith and $(i + 1)$th strands, keeping all other strands fixed. Note that the index i is the position of the strand from left to right, *not* a label for the particular strand. For n strands, we have $n - 1$ distinct generators.

Figure 6-(a) also shows the counterclockwise interchange of two strands, denoted by the operation σ_i^{-1}. The justification of the 'inverse' notation is evident in Fig. 6-(b): if we concatenate σ_i and σ_i^{-1}, then after pulling tight on the strands we find that they are disentangled. We call the braid on the right in Fig. 6-(b) the *identity braid.* In fact, the set of all braids on a given number n of strands forms a *group* in the mathematical sense: the group operation is given by concatenation of strands, the inverse by reversing the order and direction of crossings, the identity is as described above, and it is clear that concatenation is associative. This group is called B_n, the braid group on n strands, also known as the *Artin braid group* [28].

The braid group B_n is generated by the set $\{\sigma_1, \dots, \sigma_{n-1}\}$: this means that any braid in B_n can be written as a product (concatenation) of σ_i's and their inverses. The braid group is *finitely-generated*, even though it is itself infinite: only a finite number of generators give the whole group. To see that the braid group contains an infinite number of braids, simply consider σ_1^k, for k an arbitrary integer: no matter how large k gets, we always get a new braid out of this, consisting of increasingly twisted first and second strands [26].

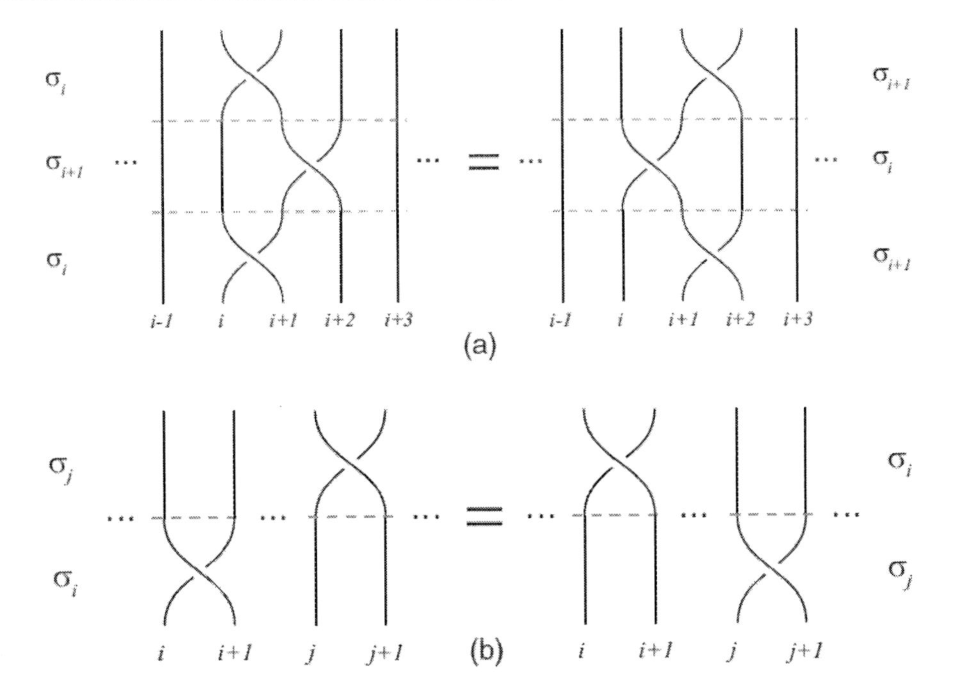

Figure 7. Braid group relations [equation (4)]: (a) relation for three adjacent strands; (b) commutation relation for generators that do not share a strand (modified and adapted from [26].

We have now passed from physical braids, as depicted in Fig. 5-(b), to *algebraic braids*. The algebraic braid corresponding to Fig. 5-(c) is $\sigma_3^{-1}\sigma_2^{-1}\sigma_3^{-1}\sigma_2\sigma_1$, where we read generators from left to right in time (beware: conventions differ). In essence, an algebraic braid is simply a sequence of generators, which may or may not come from a physical braid. How can we guarantee that physical braids and algebraic braids describe the same group? We need to be mindful of *relations* amongst the generators that arise because of physical constraints. For example, Fig. 7-(a) shows a relation amongst adjacent triplets of strands. Staring at the picture long enough, and allowing for the deformation of strands without crossing, the reader can perhaps see the that braids in Fig. 7-(a) are indeed equal. Hence, the algebraic sequence $\sigma_i\sigma_{i+1}\sigma_i$ must be equal to $\sigma_{i+1}\sigma_i\sigma_{i+1}$, if the generators are to correspond to physical braids. Another, more intuitive relation is shown in Fig. 7-(b): generators commute if they do not share a strand. In summary, we have the relations

$$\sigma_i\sigma_j\sigma_i = \sigma_j\sigma_i\sigma_j \qquad \text{if } |i-j| = 1, \tag{4a}$$

$$\sigma_i\sigma_j = \sigma_j\sigma_i \qquad \text{if } |i-j| > 1, \tag{4b}$$

amongst the generators. Artin proved in [28] the surprising fact that there are no other relations satisfied by the generators σ_i, except for those than can be derived from (4) by basic group operations (multiplication, inversion, etc.). The generators $\{\sigma_1, \ldots, \sigma_{n-1}\}$ together with the relations (4) define the *algebraic braid group*, which we also denote B_n. With these relations, the groups of physical and algebraic braids are isomorphic.

This 'braid equality' problem has seen many refinements: the original solution of

Artin [28] has computational complexity exponential in the number of generators σ_i, but modern techniques can determine equality in a time quadratic in the braid length [29, 30].

The first step in obtaining useful topological information from agent trajectories is to compute their associated braid, essentially going from the physical picture in Fig. 5-(b) to the algebraic picture in Fig. 5-(c).

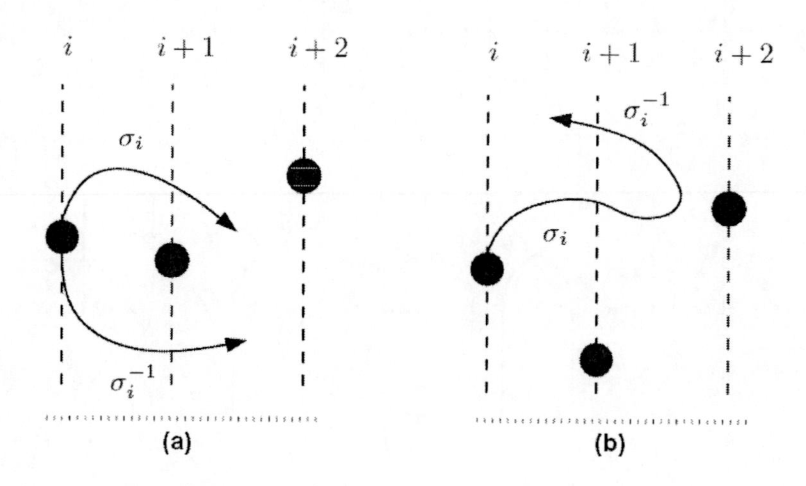

Figure 8. Detecting braids crossings: (a) Two possible agent's paths that are associated with different braid group generators; (b) Two crossings that yield no net braiding. The reference line used to detect crossings is shown dotted, and the perpendicular lines used to determine the braid generator are shown dashed (modified and adapted from [27]).

Our next goal is to map the motion of n agents onto elements of the *braid group*. First project the position of the agents onto any fixed *reference* line (which we choose to be the horizontal axis), and label the agents by $i = 1, 2, \ldots, n$ in increasing order of their projection. A crossing occurs whenever two agents interchange position on the reference line. A crossing can occur as an "over" or "under" braid, which for us means a clockwise or counterclockwise interchange. We define the braid σ_i as the clockwise interchange of the ith and $(i+1)$th agents, and σ_i^{-1} as their counterclockwise interchange, for $i = 1, \ldots, n-1$. These elementary braids are the generators of the *Artin n-braid group* [28, 31].

Assuming a crossing has occurred between the ith and $(i+1)$th agents, we need to determine if the corresponding braid generator is σ_i or σ_i^{-1}. Look at the projection of the ith and $(i+1)$th agents in the direction perpendicular to the reference line (the vertical axis in our case). If the ith agent is *above* the $(i+1)$th at the time of crossing, then the interchange involves the group generator σ_i (we define "above" as having a greater value of projection along the perpendicular direction). Conversely, if the ith agent is *below* the $(i+1)$th at the time of crossing, then the interchange involves the group generator σ_i^{-1}. Figure 8-(a) depicts these two situations.

The method just described might seem to detect spurious braids if two well-separated agents just happen to interchange position several times in a row on the reference line, as shown in Figure 8-(b). However, this would imply a sequence of σ_i and σ_i^{-1} braids, since which agent is the ith one changes at each crossing. When composed together these crossings produce no net braiding at all.

We now select a matrix representation for the generators of the braid group [32]. These matrices are given by the Burau representation[7] [33, 34, 35] of the n-braid group, which consists of $(n-1) \times (n-1)$ matrices defined by

$$[\sigma_i]_{k\ell} = \delta_{k\ell} + \delta_{k,i-1}\,\delta_{\ell i} - \delta_{k,i+1}\,\delta_{\ell i}\,, \tag{5}$$

with inverses

$$[\sigma_i^{-1}]_{k\ell} = \delta_{k\ell} - \delta_{k,i-1}\,\delta_{\ell i} + \delta_{k,i+1}\,\delta_{\ell i}\,, \tag{6}$$

where $i, k, \ell = 1, \ldots, n-1$ and we set $\delta_{k,0}$ and $\delta_{k,n}$ to zero. The determinant of each of these matrices is unity, and they satisfy the "physical braid" conditions [31]: $\sigma_i\sigma_j = \sigma_j\sigma_i$ for $|i-j| \geq 2$, and $\sigma_i\sigma_{i+1}\sigma_i = \sigma_{i+1}\sigma_i\sigma_{i+1}$. The matrices (5)–(6) can be understood as arising from the lengthening of line segments tied to the agents as the agents braid around each other [32].

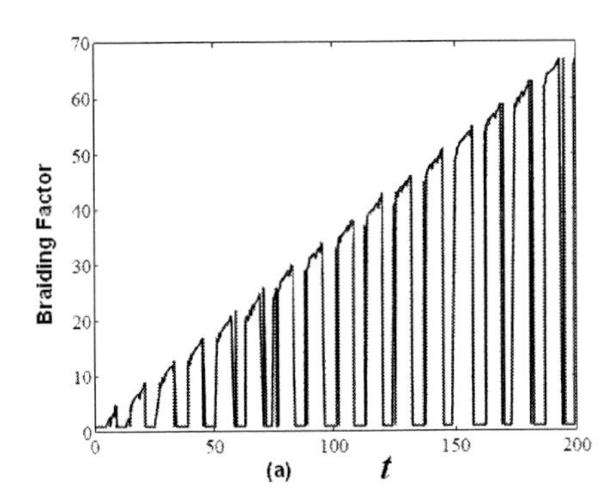

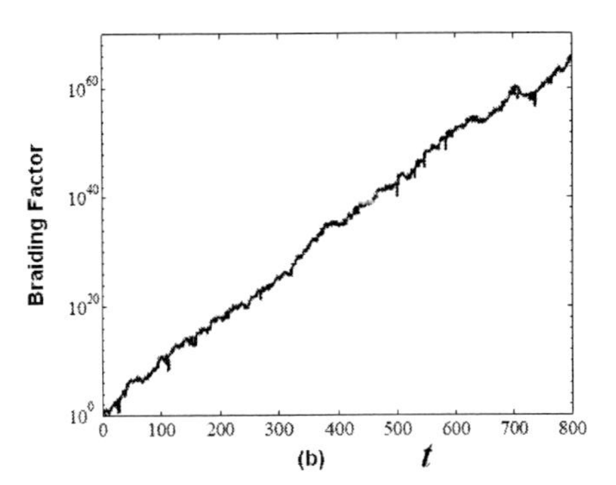

Figure 9. Braiding factor (largest eigenvalue of $\Sigma^{(N)}$) as a function of time for a triplet of agents (a) without topological chaos ($\Gamma = 0.5$); (b) with topological chaos ($\Gamma == 13$). Note that in (b) the vertical axis is logarithmic (modified and adapted from [27].

As we detect crossings, we compute the running product $\Sigma^{(N)}$ of all the braid group elements,

$$\Sigma^{(N)} = \sigma^{(N)} \cdots \sigma^{(2)}\sigma^{(1)} \tag{7}$$

where $\sigma^{(\mu)} \in \{\sigma_i\,,\, \sigma_i^{-1}\,|\, i = 1, 2, \ldots, n-1\}$ and $N(t)$ is the number of crossings detected after a time t. Now $\Sigma^{(N)}$ is the product of a sequence of (possibly random) matrices. We define the *braiding factor* to be the largest eigenvalue of $\Sigma^{(N)}$. According to Oseledec's multiplicative theorem [36], we can express the time-asymptotic exponential growth rate of the braiding factor by a (nonnegative) *Lyapunov characteristic exponent* (see Appendix), which we call the braiding exponent, defined by [27]:

$$\text{braiding exponent} = \lim_{t\to\infty} \frac{1}{t} \log|\,\text{braiding factor}\,|. \tag{8}$$

[7] More general Burau representation reads:

$$[\sigma_i]_{k\ell} = \delta_{k\ell} - \tau\,\delta_{k,i-1}\,\delta_{\ell i} - \delta_{k,i+1}\,\delta_{\ell i} - (1+\tau)\,\delta_{ki}\,\delta_{\ell i},$$

with inverses

$$[\sigma_i^{-1}]_{k\ell} = \delta_{k\ell} - \delta_{k,i-1}\,\delta_{\ell i} - \frac{1}{\tau}\,\delta_{k,i+1}\,\delta_{\ell i} - \Big(1+\frac{1}{\tau}\Big)\delta_{ki}\,\delta_{\ell i},$$

The braiding exponent is a function of the number n of braiding agents. If the exponent is positive, then we say that the sequence of braids exhibits *topological chaos*.[8]

The *braiding exponent* is our proposed measure for the *global* turbulence in multi-agents behavior dynamics. If the exponent is positive, then we say that the global turbulence occurs in a multi-agent dynamics.

5. Quantum Neural Networks (QNNs)

Now, if critical dynamics at the edge of chaos of classical adaptive systems is essentially nonlinear, then what can we say about modern *adaptive quantum systems*? While quantum mechanics is based on *superposition principle*, and thus is essentially linear, adaptation introduces critical nonlinearity. In this way, we come to the nonlinear deformation/extension of the linear *Schrödinger equation*.

An impetus to hypothesize a quantum brain model comes from the brain's necessity to unify the neuronal response into a single percept. Anatomical, neurophysiological and neuropsychological evidence, as well as brain imaging using fMRI and PET scans, show that separate functional maps exist in the biological brain to code separate features such as direction of motion, location, color and orientation. How does the brain compute all this data to have a coherent perception?

To provide a partial answer to the above question, a *quantum neural network* (QNN) (see Figure 10) has been proposed [38, 39], in which a collective response of a neuronal lattice is modeled using the Schrödinger equation:

$$\mathrm{i}\partial_t\psi(x,t) = -\frac{1}{2}\Delta\psi(x,t) + V(x)\psi(x,t), \tag{9}$$

where $\psi(x,t)$ is the wave function, or probability amplitude, associated with the quantum–mechanical system at a space-time point (x,t) and Δ is the standard *Laplacian operator*. It is shown that an external stimulus reaches each neuron in a lattice with a probability amplitude function ψ_i. Such a hypothesis would suggest that the carrier of the stimulus performs quantum computation. The collective response of all the neurons is given by the *quantum superposition equation*,

$$\psi = \sum_{i=0}^{N} c_i\psi_i = c_i\psi_i.$$

The QNN hypothesis suggests that the time evolution of the collective response $\psi = \psi(x,t)$ is described by equation (9). A neuronal lattice sets up a spatial potential field $V(x)$. A quantum process described by a quantum state ψ which mediates the collective response of a neuronal lattice evolves in the spatial potential field according to equation (9). Thus the '*classical brain*' sets up a spatio-temporal potential field $V(x,t)$ and the '*quantum brain*' is excited by this potential field to provide a collective response $\psi = \psi(x,t)$ [38, 39].

Mathematical basis for the QNN presented in Figure QuStFilt is the *nonlinear Schrödinger equation* (NLS), which gives a closed–loop feedback dynamics for the plant

[8]The braiding exponent has units of inverse time, so that if the frequency of crossings decreases then the exponent also decreases.

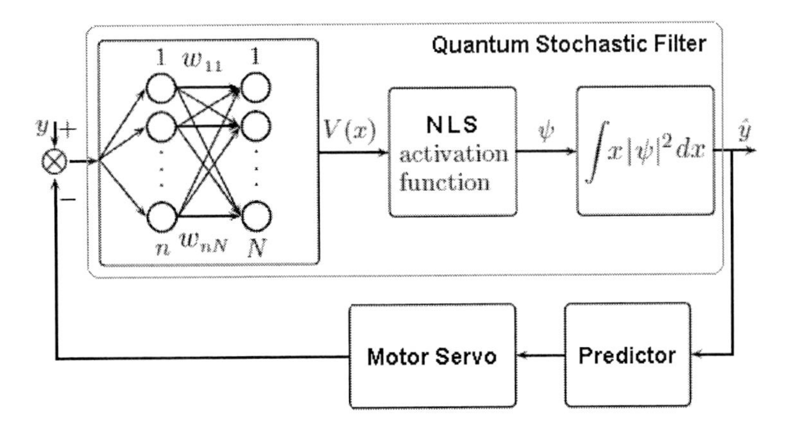

Figure 10. Quantum neural network has three blocks: quantum stochastic filter based on the *nonlinear Schrödinger equation*, Kalman–like predictor and neural motor servo, corresponding to the Kalman–like corrector (modified and adapted from [38, 40, 39]).

defined by the linear Schrödinger equation (9). Informally, QNN is represented by *adaptive NLS*, that is NLS with 'synaptic weights' w_i replacing quantum superposition constants c_i. Formally, it is the Schrödinger equation (9) with added cubic nonlinearity and synaptic weights.

Also, it was shown in [41] that quantum computation on optical modes using only beam splitters, phase shifters, photon sources and photo detectors is possible. Following this concept, [42] assumed the existence of a qubit

$$|x\rangle = \alpha|0\rangle + \beta|1\rangle, \tag{10}$$

where $|\alpha|^2 + |\beta|^2 = 1$, where the states $|0\rangle$ and $|1\rangle$ are understood as different polarization states of light.

Consider a *classical perceptron* [43], i.e., the system with n input channels $x_1, \ldots, x_n$ and one output channel y, given by

$$y = f(\sum_{j=1}^{n} w_j x_j), \tag{11}$$

where $f(\cdot)$ is the perceptron activation function and w_j are the weights tuning during learning process.

The *perceptron learning algorithm* works as follows:

1. The weights w_j are initialized to small random numbers.

2. A *pattern vector* $(x_1, \ldots, x_n)$ is presented to the perceptron and the output y generated according to the rule (11)

3. The weights are updated according to the rule

$$w_j(t+1) = w_j(t) + \eta(d - y)x_j, \tag{12}$$

   where t is discreet time, d is the desired output provided for training and $0 < \eta < 1$ is the step size.

It will be hardly possible to construct an exact analog of the nonlinear activation function f, like sigmoid and other functions of common use in neural networks, but we will show that the leaning rule of the type (12) is possible for a quantum system too.

Consider a quantum system with n inputs $|x_1\rangle, \ldots, |x_n\rangle$ of the form (10), and the output $|y\rangle$ derived by the rule

$$|y\rangle = \hat{F} \sum_{j=1}^{n} \hat{w}_j |x_j\rangle, \tag{13}$$

where $\hat{w}_j$ become 2×2 matrices acting on the basis $(|0\rangle, |1\rangle)$, combined of phase shifters $e^{i\theta}$ and beam splitters, and possibly light attenuators, $\hat{F}$ is an unknown operator that can be implemented by the network of quantum gates.

Consider the simplistic case with $\hat{F} = 1$ being the identity operator. The output of the quantum perceptron at the time t will be [42]

$$|y(t)\rangle = \sum_{j=1}^{n} \hat{w}_j(t) |x_j\rangle.$$

In analogy with classical case (12), let us provide a learning rule

$$\hat{w}_j(t+1) = \hat{w}_j(t) + \eta(|d\rangle - |y(t)\rangle)\langle x_j|, \tag{14}$$

where $|d\rangle$ is the desired output.

It can be shown that the learning rule (14) drives the quantum perceptron into desired state $|d\rangle$ used for learning. Using the rule (14) and taking the module-square difference of the real and desired outputs, we yield

$$\| |d\rangle - |y(t+1)\rangle \|^2 = \| |d\rangle - \sum_{j=1}^{n} \hat{w}_j(t+1)|x_j\rangle \|^2 = (1 - n\eta)^2 \| |d\rangle - |y(t)\rangle \|^2.$$

For small η $(0 < \eta < 1/n)$ and normalized input states $\langle x_j | x_j \rangle = 1$ the result of iteration converges to the desired state $|d\rangle$. The whole network can be then composed from the primitive elements using the standard rules of ANN architecture. For more details, see [42].

6. Adaptive Path Integral: An ∞−Dimensional QNN

6.1. Computational Partition Function

Recall that a thermodynamic *partition function* Z represents a quantity that encodes the statistical properties of a system in thermodynamic equilibrium (see, e.g. [44, 24]). It is a function of temperature and other parameters, such as the volume enclosing a gas. Other thermodynamic variables of the system, such as the total energy, free energy, entropy, and pressure, can be expressed in terms of the partition function or its partial derivatives.

A canonical ensemble is a statistical ensemble representing a probability distribution of microscopic states of the system. Its probability distribution is characterized by the proportion p_i of members of the ensemble which exhibit a measurable macroscopic state

i, where the proportion of microscopic states for each macroscopic state i is given by the Boltzmann distribution,

$$p_i = \frac{1}{Z}\mathrm{e}^{-E_i/(kT)} = \mathrm{e}^{-(E_i - A)/(kT)},$$

where E_i is the energy of state i. It can be shown that this is the distribution which is most likely, if each system in the ensemble can exchange energy with a heat bath, or alternatively with a large number of similar systems. In other words, it is the distribution which has *maximum entropy* for a given average energy $< E_i >$.

The partition function of a *canonical ensemble* is defined as a sum

$$Z(\beta) = \sum_j \mathrm{e}^{-\beta E_j}, \qquad \text{where} \qquad \beta = 1/(k_B T)$$

is the 'inverse temperature', while T is an ordinary temperature and k_B is the *Boltzmann's constant*. However, as the position x^i and momentum p_i variables of an ith particle in a system can vary continuously, the set of microstates is actually uncountable. In this case, some form of *coarse–graining* procedure must be carried out, which essentially amounts to treating two mechanical states as the same microstate if the differences in their position and momentum variables are 'small enough'. The partition function then takes the form of an integral. For instance, the partition function of a gas consisting of N molecules is proportional to the $6N$–dimensional phase–space integral,

$$Z(\beta) \sim \int_{\mathbb{R}^{6N}} d^3p_i \, d^3x^i \exp[-\beta H(p_i, x^i)],$$

where $H = H(p_i, x^i)$, $(i = 1, ..., N)$ is the classical Hamiltonian (total energy) function.

More generally, the so–called *configuration integral*, as used in probability theory, information science and dynamical systems, is an abstraction of the above definition of a partition function in statistical mechanics. It is a special case of a normalizing constant in probability theory, for the Boltzmann distribution. The partition function occurs in many problems of probability theory because, in situations where there is a natural symmetry, its associated probability measure, the *Gibbs measure*, which generalizes the notion of the canonical ensemble, has the *Markov property* [45, 46, 47].

Given a set of random variables X_i taking on values x^i, and purely potential Hamiltonian function $H(x^i)$, $(i = 1, ..., N)$, the partition function is defined as

$$Z(\beta) = \sum_{x^i} \exp\left[-\beta H(x^i)\right]. \tag{15}$$

The function H is understood to be a real-valued function on the space of states $\{X_1, X_2, \cdots\}$ while β is a real-valued free parameter (conventionally, the inverse temperature). The sum over the x^i is understood to be a sum over all possible values that the random variable X_i may take. Thus, the sum is to be replaced by an integral when the X_i are continuous, rather than discrete. Thus, one writes

$$Z(\beta) = \int dx^i \exp\left[-\beta H(x^i)\right],$$

for the case of continuously-varying random variables X_i.

The Gibbs measure of a random variable X_i having the value x^i is defined as the *probability density function* (PDF)

$$P(X_i = x^i) = \frac{1}{Z(\beta)} \exp\left[-\beta E(x^i)\right] = \frac{\exp\left[-\beta H(x^i)\right]}{\sum_{x^i} \exp\left[-\beta H(x^i)\right]},$$

where $E(x^i) = H(x^i)$ is the energy of the configuration x^i. This probability, which is now properly normalized so that $0 \leq P(x^i) \leq 1$, can be interpreted as a likelihood that a specific configuration of values x^i, $(i = 1, 2, ...N)$ occurs in the system. $P(x^i)$ is also closely related to Ω, the probability of a *random partial recursive function halting*.

As such, the partition function $Z(\beta)$ can be understood to provide the Gibbs measure on the space of states, which is the unique statistical distribution that maximizes the entropy for a fixed expectation value of the energy,

$$\langle H \rangle = -\frac{\partial \log(Z(\beta))}{\partial \beta}.$$

The associated entropy is given by

$$S = -\sum_{x^i} P(x^i) \ln P(x^i) = \beta \langle H \rangle + \log Z(\beta),$$

representing 'ignorance' + 'randomness' [45, 46, 47].

6.2. From Thermodynamics to Quantum Field Theory

In addition, the number of variables X_i in the standard partition Z need not be countable, in which case the set of coordinates $\{x^i\}$ becomes a field $\phi = \phi(x)$, so the sum is to be replaced by the *Euclidean path integral* (that is a Wick–rotated Feynman transition amplitude in imaginary time), as [24]

$$Z_{\text{Ham}}(\phi) = \int \mathcal{D}[\phi] \exp\left[-H(\phi)\right]. \tag{16}$$

More generally, in quantum field theory, instead of the field Hamiltonian $H(\phi)$ we have the action $S(\phi)$ of the theory. Both Euclidean path integral,

$$Z_{\text{Euc}}(\phi) = \int \mathcal{D}[\phi] \exp\left[-S(\phi)\right], \qquad \text{real path integral in imaginary time} \tag{17}$$

and Lorentzian one,

$$Z_{\text{Lor}}(\phi) = \int \mathcal{D}[\phi] \exp\left[\mathrm{i}S(\phi)\right], \qquad \text{complex path integral in real time} \tag{18}$$

as well as their common Lebesgue–type measure $\mathcal{D}[\phi]$, given by

$$\mathcal{D}[\phi] = \lim_{N\to\infty} \prod_{s=1}^{N} \phi_s^i, \qquad (i = 1, ..., n = \text{ number of fields}),$$

so that we can 'safely integrate over a continuous field–spectrum and sum over a discrete field spectrum' – *represent quantum field theory* (QFT) partition functions.

6.3. ∞–Dimensional QNNs

Now, a countably-infinite set of abstract synaptic weights, $\{w_s, s \in \mathbb{N}\}$, can be inserted into each of the path integrals (16), (17) and (18), transforming each of them into an ∞–dimensional quantum neural network,[9] respectively given by [45, 46, 47]:

$$\begin{aligned} Z^{\text{QNN}}_{\text{Ham}}(\phi) &= \int \mathcal{D}[w,\phi] \exp\left[-H(\phi)\right], \\ Z^{\text{QNN}}_{\text{Euc}}(\phi) &= \int \mathcal{D}[w,\phi] \exp\left[-S(\phi)\right], \\ Z^{\text{QNN}}_{\text{Lor}}(\phi) &= \int \mathcal{D}[w,\phi] \exp\left[\mathrm{i}S(\phi)\right], \end{aligned}$$

such that their *adaptive functional measure* $\mathcal{D}[w,\phi]$, given by

$$\mathcal{D}[\phi] = \lim_{N\to\infty} \prod_{s=1}^{N} w_s \phi_s^i, \qquad (i = 1, ..., n = \text{ number of fields}),$$

is trained in a neural–networks fashion, that is, by the general rule

$$new\ value(t+1) \;=\; old\ value(t) \;+\; innovation(t),$$

or more formally, by one of the standard learning rules in which the micro–time level is traversed in discrete steps, i.e., if $t = t_0, t_1, ..., t_s$ then $t+1 = t_1, t_2, ..., t_{s+1}$:

1. A *self–organized, unsupervised* (e.g., Hebbian–like [3]) learning rule:

$$w_s(t+1) = w_s(t) + \frac{\sigma}{\eta}(w_s^d(t) - w_s^a(t)), \tag{19}$$

   where $\sigma = \sigma(t)$, $\eta = \eta(t)$ denote *signal* and *noise*, respectively, while superscripts d and a denote *desired* and *achieved* micro–states, respectively; or

2. A certain form of a *supervised gradient descent learning*:

$$w_s(t+1) \;=\; w_s(t) - \eta \nabla J(t), \tag{20}$$

   where η is a small constant, called the *step size*, or the *learning rate,* and $\nabla J(n)$ denotes the gradient of the 'performance hyper–surface' at the t–th iteration; or

3. A certain form of a reward–based, *reinforcement learning* rule [48], in which system learns its optimal policy:

$$innovation(t) = |reward(t) - penalty(t)|. \tag{21}$$

---

[9] ∞ is that; ∞ is this; ∞ has come into existence from ∞. From ∞, when ∞ is taken away, ∞ remains – *Upanishads*.

7. Appendix: Lyapunov Exponents for Monitoring Computational Chaos

A branch of nonlinear and chaotic dynamics has been developed with the aim of formalizing and quantitatively characterizing the general sensitivity to initial conditions. The *largest Lyapunov exponent* λ, together with the related *Kaplan–Yorke dimension* d_{KY} and the *Kolmogorov–Sinai entropy* h_{KS} are the three indicators for measuring the *rate of error growth* produced by a dynamical system [23, 25]. The characteristic Lyapunov exponents are somehow an extension of the linear stability analysis to the case of aperiodic motions. Roughly speaking, they measure the typical rate of exponential divergence of nearby trajectories. In this sense they give information on the rate of growth of a very small error on the initial state of a system.

Consider an nD dynamical system given by the set of ODEs of the form

$$\dot{x} = f(x), \tag{22}$$

where $x = (x_1, \ldots, x_n) \in \mathbb{R}^n$ and $f : \mathbb{R}^n \to \mathbb{R}^n$. Such a system, describing spatio-temporal dynamics of human crowds, was defined in [46], from a set of *adaptive nonlinear Schrödinger equations,* using the *method of lines.* Since the right-hand-side of equation (22) does not depend on t explicitly, the system is called *autonomous*. We assume that f is smooth enough that the evolution is well defined for time intervals of arbitrary extension, and that the motion occurs in a bounded region R of the system phase space M. We intend to study the separation between two trajectories in M, $x(t)$ and $x'(t)$, starting from two close initial conditions, $x(0)$ and $x'(0) = x(0) + \delta x(0)$ in $R_0 \subset M$, respectively.

As long as the difference between the trajectories, $\delta x(t) = x'(t) - x(t)$, remains infinitesimal, it can be regarded as a vector, $z(t)$, in the tangent space $T_x M$ of M. The time evolution of $z(t)$ is given by the linearized differential equations:

$$\dot{z}_i(t) = \left.\frac{\partial f_i}{\partial x_j}\right|_{x(t)} z_j(t).$$

Under rather general hypothesis, Oseledets [36] proved that for almost all initial conditions $x(0) \in R$, there exists an orthonormal basis $\{e_i\}$ in the tangent space $T_x M$ such that, for large times,

$$z(t) = c_i e_i \exp(\lambda_i t), \tag{23}$$

where the coefficients $\{c_i\}$ depend on $z(0)$. The exponents $\lambda_1 \geq \lambda_2 \geq \cdots \geq \lambda_d$ are called *characteristic Lyapunov exponents*. If the dynamical system has an ergodic invariant measure on M, the spectrum of LEs $\{\lambda_i\}$ does not depend on the initial conditions, except for a set of measure zero with respect to the natural invariant measure.

Equation (23) describes how an nD spherical region $R = S^n \subset M$, with radius ϵ centered in $x(0)$, deforms, with time, into an ellipsoid of semi–axes $\epsilon_i(t) = \epsilon \exp(\lambda_i t)$, directed along the e_i vectors. Furthermore, for a generic small perturbation $\delta x(0)$, the distance between the reference and the perturbed trajectory behaves as

$$|\delta x(t)| \sim |\delta x(0)| \exp(\lambda_1 t) \left[1 + O\left(\exp -(\lambda_1 - \lambda_2)t\right)\right].$$

If $\lambda_1 > 0$ we have a rapid (exponential) amplification of an error on the initial condition. In such a case, the system is chaotic and, unpredictable on the long times. Indeed, if the initial error amounts to $\delta_0 = |\delta x(0)|$, and we purpose to predict the states of the system with a certain tolerance Δ, then the prediction is reliable just up to a *predictability time* given by

$$T_p \sim \frac{1}{\lambda_1} \ln\left(\frac{\Delta}{\delta_0}\right).$$

This equation shows that T_p is basically determined by the *positive leading Lyapunov exponent*, since its dependence on δ_0 and Δ is logarithmically weak. Because of its preeminent role, λ_1 is often referred as 'the leading positive Lyapunov exponent', and denoted by λ.

Therefore, Lyapunov exponents are average rates of expansion or contraction along the principal axes. For the ith principal axis, the corresponding Lyapunov exponent is defined as

$$\lambda_i = \lim_{t\to\infty} \{(1/t) \ln[L_i(t)/L_i(0)]\}, \tag{24}$$

where $L_i(t)$ is the radius of the ellipsoid along the ith principal axis at time t.

An initial volume V_0 of the phase–space region R_0 evolves on average as

$$V(t) = V_0 e^{(\lambda_1+\lambda_2+\cdots+\lambda_{2n})t}, \tag{25}$$

and therefore the rate of change of $V(t)$ is simply

$$\dot{V}(t) = \sum_{i=1}^{2n} \lambda_i V(t).$$

In the case of a 2D phase area A, evolving as $A(t) = A_0 e^{(\lambda_1+\lambda_2)t}$, a *Lyapunov dimension* d_L is defined as

$$d_L = \lim_{\epsilon\to 0} \left[\frac{d(\ln(N(\epsilon)))}{d(\ln(1/\epsilon))}\right],$$

where $N(\epsilon)$ is the number of squares with sides of length ϵ required to cover $A(t)$, and d represents an ordinary *capacity dimension*,

$$d_c = \lim_{\epsilon\to 0} \left(\frac{\ln N}{\ln(1/\epsilon)}\right).$$

Lyapunov dimension can be extended to the case of nD phase–space by means of the *Kaplan–Yorke dimension* [49, 50, 51]) as

$$d_{KY} = j + \frac{\lambda_1 + \lambda_2 + \cdots + \lambda_j}{|\lambda_{j+1}|},$$

where the λ_i are ordered (λ_1 being the largest) and j is the index of the smallest nonnegative Lyapunov exponent.

On the other hand, a state, initially determined with an error $\delta x(0)$, after a time enough larger than $1/\lambda$, may be found almost everywhere in the region of motion $R \in M$. In this respect, the *Kolmogorov–Sinai* (KS) *entropy*, h_{KS}, supplies a more refined information. The error on the initial state is due to the maximal resolution we use for observing the system. For simplicity, let us assume the same resolution ϵ for each degree of freedom. We build a partition of the phase space M with cells of volume ϵ^d, so that the state of the system at $t = t_0$ is found in a region R_0 of volume $V_0 = \epsilon^d$ around $x(t_0)$. Now we consider the trajectories starting from V_0 at t_0 and sampled at discrete times $t_j = j\,\tau$ ($j = 1, 2, 3, \ldots, t$). Since we are considering motions that evolve in a bounded region $R \subset M$, all the trajectories visit a finite number of different cells, each one identified by a symbol. In this way a unique sequence of symbols $\{s(0), s(1), s(2), \ldots\}$ is associated with a given trajectory $x(t)$. In a chaotic system, although each evolution $x(t)$ is univocally determined by $x(t_0)$, a great number of different symbolic sequences originates by the same initial cell, because of the divergence of nearby trajectories. The total number of the admissible symbolic sequences, $\widetilde{N}(\epsilon, t)$, increases exponentially with a rate given by the topological entropy

$$h_T = \lim_{\epsilon \to 0} \lim_{t \to \infty} \frac{1}{t} \ln \widetilde{N}(\epsilon, t)\,.$$

However, if we consider only the number of sequences $N_{eff}(\epsilon, t) \leq \widetilde{N}(\epsilon, t)$ which appear with very high probability in the long time limit – those that can be numerically or experimentally detected and that are associated with the natural measure – we arrive at a more physical quantity called the Kolmogorov–Sinai (or metric) entropy, which is the key entropy notion in ergodic theory [52]:

$$h_{KS} = \lim_{\epsilon \to 0} \lim_{t \to \infty} \frac{1}{t} \ln N_{eff}(\epsilon, t) \leq h_T. \tag{26}$$

h_{KS} quantifies the long time exponential rate of growth of the number of the effective coarse-grained trajectories of a system. This suggests a link with information theory where the Shannon entropy measures the mean asymptotic growth of the number of the typical sequences – the ensemble of which has probability almost one – emitted by a source.

We may wonder what is the number of cells where, at a time $t > t_0$, the points that evolved from R_0 can be found, i.e., we wish to know how big is the coarse–grained volume $V(\epsilon, t)$, occupied by the states evolved from the volume V_0 of the region R_0, if the minimum volume we can observe is $V_{min} = \epsilon^d$. As stated above (25), we have

$$V(t) \sim V_0 \exp(t \sum_{i=1}^{d} \lambda_i).$$

However, this is true only in the limit $\epsilon \to 0$. In this (unrealistic) limit, $V(t) = V_0$ for a conservative system (where $\sum_{i=1}^{d} \lambda_i = 0$) and $V(t) < V_0$ for a dissipative system (where $\sum_{i=1}^{d} \lambda_i < 0$). As a consequence of limited resolution power, in the evolution of the volume $V_0 = \epsilon^d$ the effect of the contracting directions (associated with the negative Lyapunov exponents) is completely lost. We can experience only the effect of the expanding directions, associated with the positive Lyapunov exponents. As a consequence, in the typical case, the coarse grained volume behaves as

$$V(\epsilon, t) \sim V_0\, \mathrm{e}^{(\sum_{\lambda_i > 0} \lambda_i)\, t},$$

when V_0 is small enough. Since $N_{eff}(\epsilon, t) \propto V(\epsilon, t)/V_0$, one has: $h_{KS} = \sum_{\lambda_i > 0} \lambda_i$. This argument can be made more rigorous with a proper mathematical definition of the metric entropy. In this case one derives the Pesin relation [53, 52]: $h_{KS} \leq \sum_{\lambda_i>0} \lambda_i$. Because of its relation with the Lyapunov exponents, or by the definition (26), it is clear that also h_{KS} is a fine-grained and global characterization of a dynamical system.

The metric entropy is an invariant characteristic quantity of a dynamical system, i.e., given two systems with invariant measures, their KS–entropies exist and they are equal iff the systems are isomorphic [54].

Finally, the *topological entropy* on the manifold M equals the supremum of the Kolmogorov-Sinai entropies,

$$h(u) = \sup\{h_{KS}(u) = h_\mu(u) : \mu \in P_u(M)\},$$

where $u : M \to M$ is a continuous map on M, and μ ranges over all $u-$invariant (Borel) probability measures on M. Dynamical systems of positive topological entropy are often considered topologically chaotic.

References

[1] V. Ivancevic, T. Ivancevic, *Quantum Neural Computation,* Springer, (2009)

[2] Haykin, S.S.: *Neural Networks: A Comprehensive Foundation* (2nd ed). Prentice Hall, NJ, (1998)

[3] Hebb, D.O.: *The Organization of Behavior*. Wiley, New York, (1949)

[4] Kosko, B.: *Neural Networks, Fuzzy Systems, A Dynamical Systems Approach to Machine Intelligence.* Prentice–Hall, New York, (1992)

[5] Ivancevic, V., Ivancevic, T.: *Neuro-Fuzzy Associative Machinery for Comprehensive Brain and Cognition Modelling.* Springer, Berlin, (2007)

[6] Ivancevic, V., Ivancevic, T.: *Computational Mind: A Complex Dynamics Perspective.* Springer, Berlin, (2007)

[7] Hopfield, J.J.: Neurons with graded response have collective computational properties like those of two–state neurons. *Proc. Natl. Acad. Sci. USA,* **81**, 3088–3092, (1984)

[8] Hecht-Nielsen, R.: Counterpropagation networks. *Applied Optics,* **26**(23), 4979–4984, (1987)

[9] Hecht-Nielsen, R.: *NeuroComputing.* Addison–Wesley, Reading, (1990)

[10] Cohen, M.A., Grossberg, S.: Absolute stability of global pattern formation, parallel memory storage by competitive neural networks. *IEEE Trans. Syst., Man, Cybern.,* **13**(5), 815–826, (1983)

[11] Carpenter, G.A., Grossberg, S.: Adaptive Resonance Theory. In M.A. Arbib (ed.) *The Handbook of Brain Theory*, Neural Networks, Second Edition, MIT Press, Cambridge, MA, 87–90, (2003)

[12] Hodgkin, A.L., Huxley, A.F.: A quantitative description of membrane current and application to conduction and excitation in nerve. *J. Physiol.,* **117**, 500-544, (1952)

[13] Hodgkin, A.L.: *The Conduction of the Nervous Impulse.* Liverpool Univ. Press, Liverpool, (1964)

[14] Feynman, R.P., Simulating physics with computers. *Int. J. Th. Phys.* **21**(6/7), 467–488, (1982)

[15] Benioff, P.A., Quantum mechanical Hamiltonian models of Turing machines. *J. Stat. Phys.* **29**(3), 515–546, (1982)

[16] Deutsch, D., Quantum theory, the Church-Turing principle and the universal quantum computer. *Proc. Roy. Soc. London A* **400**, 97–117, (1985)

[17] Bernstein, E., Vazirani, U., Quantum complexity theory. *SIAM J. Comput.* **26**(5), 1411–1473, (1997)

[18] Yao, A., Quantum circuit complexity. In *Proc 34th IEEE Symposium on Foundations of computer science,* 352–361, (1993)

[19] Gershenfeld, N., Chuang, I.L., *Quantum Computing with Molecules.* Scientific American, June, (1998)

[20] Craddock, T.J.A., Tuszynski, J.A., On the Role of the Microtubules in Cognitive Brain Functions. *NeuroQuant.* **5**(1), 32–57, (2007)

[21] Collins, G.P., *Computing with Quantum Knots.* Scientific American, April, (2006)

[22] Nayak, C., Simon, S.H., Stern, A., Freedman, M., Sarma, S.D., Non-Abelian Anyons and Topological Quantum Computation. *Rev. Mod. Phys.* **80**, 1083 (77 pages), (2008)

[23] Ivancevic, V., Ivancevic, T., *High–Dimensional Chaotic and Attractor Systems.* Springer, Berlin, (2007)

[24] Ivancevic, V., Ivancevic, T. Complex Nonlinearity: Chaos, Phase Transitions, *Topology Change and Path Integrals*. Springer, Berlin, (2008)

[25] Ivancevic, T., Jain, L., Pattison, J., Hariz, A. Nonlinear Dynamics and Chaos Methods in Neurodynamics and Complex Data *Analysis. Nonl. Dyn.* **56**(1-2), 23-44, (2009)

[26] J-L. Thiffeault, Braids of entangled particles trajectories. *CHAOS* **20**, 017516, (2010)

[27] J-L. Thiffeault, Measuring topological chaos. *Phys. Rev. Let.* **94**, 084502, (2005)

[28] E. Artin, Theory of Braids. *Ann. Math.* **48**, 101-126, (1947)

[29] J.S. Birman, K.H. Ko, S.J. Lee, A new approach to the word and conjugacy problem in the braid groups. *Adv. Math.* **139**, 322353, (1998)

[30] I. Dynnikov, B. Wiest, On the Complexity of Braids. *J. Eur. Math. Soc.* **9**, 801840, (2007)

[31] K. Murasugi, *Knot theory and its applications.* Birkhäuser, Boston, (1996)

[32] P.L. Boyland, H. Aref, M.A. Stremler, Topological fluid mechanics of stirring. *J. Fluid Mech.* **403**, 277, (2000)

[33] W. Burau, Über Zopfgruppen und gleichsinnig verdrilte Verkettungen. *Abh. Math. Sem. Hanischen Univ.* **11**, 171, (1936)

[34] V.F.R. Jones, Hecke algebra representations of Braid groups and link polynomials. *Ann. Math.* **126**, 335-388, (1987)

[35] *Burau Representation, Wolfram MathWorld,* (2009) http://mathworld.wolfram.com/BurauRepresentation.html

[36] Oseledets, V.I. A Multiplicative Ergodic Theorem: Characteristic Lyapunov Exponents of Dynamical Systems. *Trans. Moscow Math. Soc.*, **19**, 197–231, (1968)

[37] Bertschinger, N., Natschläger, T.: *Real-Time Computation Neural Networks. Neu. Comp.* **16**, 1413-1436, (2004)

[38] Behera, L., Kar, I., Elitzur, A.C.: Recurrent Quantum Neural Network Model to Describe Eye Tracking of Moving Target, *Found. Phys. Let.* **18**(4), 357-370, (2005)

[39] Behera, L., Kar, I., Elitzur, A.C.: *Recurrent Quantum Neural Network and its Applications,* Chapter 9 in The Emerging Physics of Consciousness, J.A. Tuszynski (ed.) Springer, Berlin, (2006)

[40] Behera, L., Kar, I.: Quantum Stochastic Filtering. In: *Proc. IEEE Int. Conf. SMC* **3**, 2161–2167, (2005)

[41] Knill, E., Laflamme, R., Milburn, G.J.: A scheme for efficient quantum computation with linear optics, *Nature* **409**, 46-57, (2001)

[42] Altaisky, M.V.: *Quantum neural network.* arXiv:quant-ph/0107012, (2001)

[43] Minsky, M., Papert, S.: *Perceptrons.* MIT Press, Cambridge, MA, (1969)

[44] Ivancevic, V., Ivancevic, T.: *Natural Biodynamics.* World Scientific, Singapore, (2006)

[45] V. Ivancevic, D. Reid, E. Aidman, Crowd behavior dynamics: entropic path-integral model. *Nonl. Dyn.* **59**, 351-373, (2010)

[46] V. Ivancevic, D. Reid, Dynamics of Confined Crowds Modelled using Entropic Stochastic Resonance and Quantum Neural Networks. *Int. J. Intel. Defence Sup. Sys.* **2**(4), 269-289, (2009)

[47] V. Ivancevic, D. Reid, Crowd behavior dynamics: entropic path-integral model. Nonl. Dyn. Entropic geometry of crowd dynamics. *A Chapter in Nonlear Dynamics* (T. Evancs, Ed.), Intech, Vienna, (2010)

[48] Sutton, R.S., Barto, A.G.: *Reinforcement Learning: An Introduction.* MIT Press, Cambridge, MA, (1998)

[49] Kaplan, J.L., Yorke, J.A. Numerical Solution of a Generalized Eigenvalue Problem for Even Mapping. Peitgen, H.O., Walther, H.O. (eds.). Functional Differential Equations and Approximations of Fixed Points, *Lecture Notes in Mathematics,* **730**, Springer, Berlin, 228–256, (1979)

[50] Yorke, J.A., Alligood, K., Sauer, T. *Chaos: An Introduction to Dynamical Systems.* Springer, New York, (1996)

[51] Ott, E., Grebogi, C., Yorke, J.A. Controlling chaos. *Phys. Rev. Lett.* **64**, 1196–1199, (1990)

[52] Eckmann, J.P., Ruelle, D. Ergodic theory of chaos and strange attractors. *Rev. Mod. Phys.* **57**, 617–630, (1985)

[53] Pesin, Ya.B. Lyapunov Characteristic Exponents and Smooth Ergodic Theory. *Russ. Math. Surveys,* **32**(4), 55–114, (1977)

[54] Billingsley, P. *Ergodic theory and information*. Wiley, New York, (1965)

In: Computational Engineering
Editors: J. E. Browning and A. K. McMann
ISBN: 978-1-61122-806-9

Chapter 4

LINEAR VERSUS NONLINEAR HUMAN OPERATOR MODELING

Tijana T. Ivancevic**, *Bojan N. Jovanovic, Sasha A. Jovanovic, Leon Lukman, Alexandar Lukman and Milka Djukic
Society for Nonlinear Dynamics in Human Factors, Adelaide, Australia
and CITECH Research IP Pty Ltd, Adelaide, Australia

Abstract

The motivation behind mathematically modeling the *human operator* is to help explain the response characteristics of the complex dynamical system including the human manual controller. In this paper, we present two approaches to human operator modeling: classical linear control approach and modern nonlinear control approach. The latter one is formalized using both fixed and adaptive Lie-Derivative based controllers.

Keywords: Human operator, linear control, nonlinear control, Lie derivative operator

1. Introduction

Despite the increasing trend toward automation, robotics and artificial intelligence (AI) in many environments, the *human operator* will probably continue for some time to be integrally involved in the control and regulation of various machines (e.g., missile–launchers, ground vehicles, watercrafts, submarines, spacecrafts, helicopters, jet fighters, etc.). A typical manual control task is the task in which control of these machines is accomplished by *manipulation of the hands or fingers* [1]. As human–computer interfaces evolve, interaction techniques increasingly involve a much more continuous form of interaction with the user, over both human–to–computer (input) and computer–to–human (output) channels. Such interaction could involve gestures, speech and animation in addition to more 'conventional' interaction via mouse, joystick and keyboard. This poses a problem for the design of interactive systems as it becomes increasingly necessary to consider interactions occurring over an interval, in continuous time.

*E-mail address: tijana.ivancevic@alumni.adelaide.edu.au. (Corresponding author)

The so–called *manual control theory* developed out of the efforts of feedback control engineers during and after the World War II, who required models of human performance for continuous military tasks, such as tracking with anti–aircraft guns [2]. This seems to be an area worth exploring, firstly since it is generally concerned with systems which are controlled in continuous time by the user, although discrete time analogues of the various models exist. Secondly, it is an approach which models both system and user and hence is compatible with research efforts on 'synthetic' models, in which aspects of both system and user are specified within the same framework. Thirdly, it is an approach where continuous mathematics is used to describe functions of time. Finally, it is a theory which has been validated with respect to experimental data and applied extensively within the military domains such as avionics.

The premise of manual control theory is that for certain tasks, the performance of the human operator can be well approximated by a describing function, much as an inanimate controller would be. Hence, in the literature frequency domain representations of behavior in continuous time are applied. Two of the main classes of system modelled by the theory are *compensatory* and *pursuit* systems. A system where only the error signal is available to the human operator is a compensatory system. A system where both the target and current output are available is called a pursuit system. In many pursuit systems the user can also see a portion of the input in advance; such tasks are called *preview tasks* [3].

A simple and widely used model is the 'crossover model' [9], which has two main parameters, a *gain* K and a *time delay* τ, given by the transfer function in the Laplace transform s domain

$$H = K\frac{\mathrm{e}^{-\tau s}}{s}.$$

Even with this simple model we can investigate some quite interesting phenomena. For example consider a compensatory system with a certain delay, if we have a low gain, then the system will move only slowly towards the target, and hence will seem sluggish. An expanded version of the crossover model is given by the transfer function [1]

$$H = K\frac{(T_L s + 1)\,\mathrm{e}^{-(\tau s + \alpha/s)}}{(T_I s + 1)(T_N s + 1)},$$

where T_L and T_I are the lead and lag constants (which describe the *equalization* of the human operator), while the first–order lag $(T_N S + 1)$ approximates the neuromuscular lag of the hand and arm. The expanded term α/s in the time delay accounts for the 'phase drop', i.e., increased lags observed at very low frequency [4].

Alternatively if the gain K is very high, then the system is very likely to overshoot the target, requiring an adjustment in the opposite direction, which may in turn overshoot, and so on. This is known as 'oscillatory behavior'. Many more detailed models have also been developed; there are 'anthropomorphic models', which have a cognitive or physiological basis. For example the 'structural model' attempts to reflect the structure of the human, with central nervous system, neuromuscular and vestibular components [3]. Alternatively there is the 'optimal control modeling' approach, where algorithmic models which very closely match empirical data are used, but which do not have any direct relationship or explanation in terms of human neural and cognitive architecture [10]. In this model, an operator is assumed to perceive a vector of displayed quantities and must exercise control to minimize a *cost functional* given by [1]

$$J = E\{\lim_{T\to\infty} \frac{1}{T} \int_0^T [q_i y_i^2(t) + \sum_i (r_i u^2(t) + g_i \dot{u}^2(t))] dt\},$$

which means that the operator will attempt to minimize the expected value E of some weighted combination of squared display error y, squared control displacement u and squared control velocity $\dot{u}$. The relative values of the weighting constants q_i, r_i, g_i will depend upon the relative importance of control precision, control effort and fuel expenditure.

In the case of manual control of a vehicle, this modeling yields the 'closed–loop' or 'operator–vehicle' dynamics. A quantitative explanation of this closed–loop behavior is necessary to summarize operator behavioral data, to understand operator control actions, and to predict the operator–vehicle dynamic characteristics. For these reasons, control engineering methodologies are applied to modeling human operators. These 'control theoretic' models primarily attempt to represent the operator's control behavior, not the physiological and psychological structure of the operator [6, 7]. These models 'gain in acceptability' if they can identify features of these structures, 'although they cannot be rejected' for failing to do so [8].

One broad division of human operator models is whether they simulated a continuous or discontinuous operator control strategy. Significant success has been achieved in modeling human operators performing compensatory and pursuit tracking tasks by employing continuous, quasi–linear operator models. Examples of these include the crossover optimal control models mentioned above.

Discontinuous input behavior is often observed during manual control of large amplitude and acquisition tasks [9, 11, 12, 13]. These discontinuous human operator responses are usually associated with *precognitive human control behavior* [9, 14]. Discontinuous control strategies have been previously described by 'bang–bang' or relay control techniques. In [15], the authors highlighted operator's preference for this type of relay control strategy in a study that compared controlling high–order system plants with a proportional verses a relay control stick. By allowing the operator to generate a sharper step input, the relay control stick improved the operators' performance by up to 50 percent. These authors hypothesized that when a human controls a high–order plant, the operator must consider the error of the system to be dependent upon the integral of the control input. Pulse and step inputs would reduce the integration requirements on the operator and should make the system error response more predictable to the operator.

Although operators may employ a *bang–bang control* strategy, they often impose an internal limit on the magnitude of control inputs. This internal limit is typically less than the full control authority available [9]. Some authors [16] hypothesized that this behavior is due to the operator's recognition of their own reaction time delay. The operator must tradeoff the cost of a switching time error with the cost of limiting the velocity of the output to a value less than the maximum.

A significant amount of research during the 1960's and 1970's examined discontinuous input behavior by human operators and developed models to emulate it [14, 17, 18, 19, 20, 21, 22, 23, 24]. Good summaries of these efforts can be found in [25], [11], [9] and [6, 7]. All of these efforts employed some type of *relay* element to model the discontinuous input behavior. During the 1980's and 1990's, pilot models were developed that included switching or discrete changes in pilot behavior [26, 27, 28, 29, 12, 13].

Recently, the so-called 'variable structure control' techniques were applied to model human operator behavior during acquisition tasks [6, 7]. The result was a coupled, multi–input model replicating the discontinuous control strategy. In this formulation, a switching surface was the mathematical representation of the human operator's control strategy. The performance of the variable strategy model was evaluated by considering the longitudinal control of an aircraft during the visual landing task.

In this paper, we present two approaches to human operator modeling: classical linear control approach and modern nonlinear Lie-Derivative based control approach.

2. Classical Control Theory versus Nonlinear Dynamics and Control

In this section we review classical feedback control theory (see e.g., [30, 4, 31]) and contrast it with nonlinear and stochastic dynamics (see e.g., [32, 33, 34]).

2.1. Basics of Kalman's Linear State–Space Theory

Linear multiple input–multiple output (MIMO) control systems can always be put into Kalman canonical state–space form of order n, with m inputs and k outputs. In the case of *continual time* systems we have state and output equation of the form

$$\begin{aligned} d\mathbf{x}/dt &= \mathbf{A}(t)\,\mathbf{x}(t) + \mathbf{B}(t)\,\mathbf{u}(t), \\ \mathbf{y}(t) &= \mathbf{C}(t)\,\mathbf{x}(t) + \mathbf{D}(t)\,\mathbf{u}(t), \end{aligned} \tag{1}$$

while in case of *discrete time* systems we have state and output equation of the form

$$\begin{aligned} \mathbf{x}(n+1) &= \mathbf{A}(n)\,\mathbf{x}(n) + \mathbf{B}(n)\,\mathbf{u}(n), \\ \mathbf{y}(n) &= \mathbf{C}(n)\,\mathbf{x}(n) + \mathbf{D}(n)\,\mathbf{u}(n). \end{aligned} \tag{2}$$

Both in (1) and in (2) the variables have the following meaning:

$\mathbf{x}(t) \in \mathbb{X}$ is an $n-$vector of *state variables* belonging to the *state space* $\mathbb{X} \subset \mathbb{R}^n$;
$\mathbf{u}(t) \in \mathbb{U}$ is an $m-$vector of *inputs* belonging to the *input space* $\mathbb{U} \subset \mathbb{R}^m$;
$\mathbf{y}(t) \in \mathbb{Y}$ is a $k-$vector of *outputs* belonging to the *output space* $\mathbb{Y} \subset \mathbb{R}^k$;
$\mathbf{A}(t) : \mathbb{X} \to \mathbb{X}$ is an $n \times n$ matrix of *state dynamics*;
$\mathbf{B}(t) : \mathbb{U} \to \mathbb{X}$ is an $n \times m$ matrix of *input map*;
$\mathbf{C}(t) : \mathbb{X} \to \mathbb{Y}$ is an $k \times n$ matrix of *output map*;
$\mathbf{D}(t) : \mathbb{U} \to \mathbb{Y}$ is an $k \times m$ matrix of *input–output transform*.

Input $\mathbf{u}(t) \in \mathbb{U}$ can be empirically determined by trial and error; it is properly defined by optimization process called *Kalman regulator*, or more generally (in the presence of noise), by *Kalman filter* (even better, *extended Kalman filter* to deal with stochastic nonlinearities).

2.2. Linear Stationary Systems and Operators

The most common special case of the general Kalman model (1), with constant state, input and output matrices (and relaxed boldface vector–matrix notation), is the so–called *stationary linear model*

$$\dot{x} = Ax + Bu, \qquad y = Cx. \tag{3}$$

The stationary linear system (3) defines a variety of operators, in particular those related to the following problems:

1. regulators,
2. end point controls,
3. servomechanisms, and
4. repetitive modes (see [39]).

2.2.1. Regulator Problem and the Steady State Operator

Consider a variable, or set of variables, associated with a dynamical system. They are to be maintained at some desired values in the face of changing circumstances. There exist a second set of parameters that can be adjusted so as to achieve the desired regulation. The effecting variables are usually called *inputs* and the affected variables called *outputs*. Specific examples include the regulation of the thrust of a jet engine by controlling the flow of fuel, as well as the regulation of the oxygen content of the blood using the respiratory rate.

Now, there is the steady state operator of particular relevance for the regulator problem. It is

$$y_\infty = -CA^{-1}Bu_\infty,$$

which describes the map from constant values of u to the equilibrium value of y. It is defined whenever A is invertible but the steady state value will only be achieved by a real system if, in addition, the eigenvalues of A have negative real parts. Only when the rank of $CA^{-1}B$ equals the dimension of y can we steer y to an arbitrary steady state value and hold it there with a constant u. A nonlinear version of this problem plays a central role in robotics where it is called the *inverse kinematics problem* (see, e.g., [40]).

2.2.2. End Point Control Problem and the Adjustment Operator

Here we have inputs, outputs and trajectories. In this case the shape of the trajectory is not of great concern but rather it is the end point that is of primary importance. Standard examples include rendezvous problems such as one has in space exploration.

Now, the operator of relevance for the end point control problem, is the operator

$$x(T) = \int_0^T \exp[A(T-\sigma)]\, Bu(\sigma)\, d\sigma.$$

If we consider this to define a map from the mD L_2 space $L_2^m[0,T]$ (where u takes on its values) into $\mathbb{R}^m$ then, if it is an onto map, it has a Moore–Penrose (least squares) inverse

$$u(\sigma) = B^T \exp[A^T(T-\sigma)]\,(W[0,T])^{-1}\,(x(T) - \exp(AT)\,x(0)),$$

with the symmetric positive definite matrix W, the *controllability Gramian*, being given by

$$W[0,T] = \int_0^T \exp[A(T-\sigma)]\,BB^T \exp[A^T(T-\sigma)]\,d\sigma.$$

2.2.3. Servomechanism Problem and the Corresponding Operator

Here we have inputs, outputs and trajectories, as above, and an associated dynamical system. In this case, however, it is desired to cause the outputs to follow a trajectory specified by the input. For example, the control of an airplane so that it will travel along the flight path specified by the flight controller.

Now, because we have assumed that A, B and C are constant

$$y(t) = C\exp(At)\,x(0) + \int_0^t C\exp[A(T-\tau)]\,Bu(\tau)\,d\tau,$$

and, as usual, the Laplace transform can be used to convert convolution to multiplication. This brings out the significance of the Laplace transform pair

$$C\exp(At)B \Longleftrightarrow C(Is-A)^{-1}B \tag{4}$$

as a means of characterizing the input–output map of a linear model with constant coefficients.

2.2.4. Repetitive Mode Problem and the Corresponding Operator

Here again one has some variable, or set of variables, associated with a dynamical system and some inputs which influence its evolution. The task has elements which are repetitive and are to be done efficiently. Examples from biology include the control of respiratory processes, control of the pumping action of the heart, control of successive trials in practicing a athletic event.

The relevant operator is similar to the servomechanism operator, however the constraint that u and x are periodic means that the relevant diagonalization is provided by Fourier series, rather than the Laplace transform. Thus, in the Fourier domain, we are interested in a set of complex matrices

$$G(iw_i) = C(iw_i - A)^{-1}B, \qquad w_i = 0, w_0, 2w_0, \ldots$$

More general, but still deterministic, models of the input–state–output relation are afforded by the *nonlinear affine model* (see, e.g., [41])

$$\begin{aligned} \dot{x}(t) &= f(x(t)) + g(x(t))\,u(t), \\ y(t) &= h(x(t)); \end{aligned}$$

and the still more general *fully nonlinear model*

$$\begin{aligned}\dot{x}(t) &= f(x(t), u(t)),\\ y(t) &= h(x(t)).\end{aligned}$$

2.2.5. Feedback Changes the Operator

No idea is more central to automatic control than the idea of feedback. When an input is altered on the basis of the difference between the actual output of the system and the desired output, the system is said to involve *feedback.* Man made systems are often constructed by starting with a basic element such as a motor, a burner, a grinder, etc. and then adding sensors and the hardware necessary to use the measurement generated by the sensors to regulate the performance of the basic element. This is the essence of *feedback control.* Feedback is often contrasted with open loop systems in which the inputs to the basic element is determined without reference to any measurement of the trajectories. When the word feedback is used to describe naturally occurring systems, it is usually implicit that the behavior of the system can best be explained by pretending that it was designed as one sees man made systems being designed [39].

In the context of linear systems, the effect of feedback is easily described. If we start with the stationary linear system (3) with u being the controls and y being the measured quantities, then the effect of feedback is to replace u by $u - Ky$ with K being a matrix of feedback gains. The closed–loop equations are then

$$\dot{x} = (A - BKC)\,x + Bu, \qquad y = Cx.$$

Expressed in terms of the Laplace transform pairs (4), feedback effects the transformation

$$\left(C\exp(At)B; C(Is - A)^{-1}B\right) \longmapsto C\exp(A - BKC)^{t}B; C(Is - A + BKC)^{-1}B.$$

Using such a transformation, it is possible to alter the dynamics of a system in a significant way. The modifications one can effect by feedback include influencing the location of the eigenvalues and consequently the stability of the system. In fact, if K is m by p and if we wish to select a gain matrix K so that $A - BKC$ has eigenvalues $\lambda_1, \lambda_2, ..., \lambda_n$, it is necessary to insure that

$$\det\begin{pmatrix} C(I\lambda_1 - A)^{-1}B & -I \\ I & K \end{pmatrix} = 0, \qquad i = 1, 2, ..., n.$$

Now, if CB is invertible then we can use the relationship $C\dot{x} = CAx + CBu$ together with $y = Cx$ to write $\dot{y} = CAx + CBu$. This lets us solve for u and recast the system as

$$\begin{aligned}\dot{x} &= (A - B(CB)^{-1}CA)\,x + B(CB)^{-1}\dot{y},\\ u &= (CB)^{-1}\dot{y} - (CB)^{-1}CA\,x.\end{aligned}$$

Here we have a set of equations in which the roles of u and y are reversed. They show how a choice of y determines x and how x determines u [39].

2.3. Stability and Boundedness

Let a time–varying dynamical system may be expressed as

$$\dot{x}(t) = f(t, x(t)) \tag{5}$$

where $x \in \mathbb{R}^n$ is an nD vector and $f : \mathbb{R}^+ \times D \to \mathbb{R}^n$ with $D = \mathbb{R}^n$ or $D = B_h$ for some $h > 0$, where $B_h = \{x \in \mathbb{R}^n : |x| < h\}$ is a ball centered at the origin with a radius of h. If $D = \mathbb{R}^n$ then we say that the dynamics of the system are defined *globally*, whereas if $D = B_h$ they are only defined *locally*. We do not consider systems whose dynamics are defined over disjoint subspaces of R. It is assumed that $f(t, x)$ is piecemeal continuous in t and Lipschitz in x for existence and uniqueness of state solutions. As an example, the linear system $\dot{x}(t) = Ax(t)$ fits the form of (5) with $D = \mathbb{R}^n$ [38].

Assume that for every x_0 the initial value problem

$$\dot{x}(t) = f(t, x(t)), \qquad x(t_0) = x_0,$$

possesses a unique solution $x(t, t_0, x_0)$; it is called a solution to (5) if $x(t, t_0, x_0) = x_0$ and $\frac{d}{dt}x(t, t_0, x_0) = f(t, x(t, t_0, x_0))$ [38].

A point $x_e \in \mathbb{R}^n$ is called an *equilibrium point* of (5) if $f(t, x_e) = 0$ for all $t \geq 0$. An equilibrium point x_e is called an *isolated equilibrium point* if there exists an $\rho > 0$ such that the ball around x_e, $B_\rho(x_e) = \{x \in \mathbb{R}^n : |x - x_e| < \rho\}$, contains no other equilibrium points besides x_e [38].

The equilibrium $x_e = 0$ of (5) is said to be *stable in the sense of Lyapunov* if for every $\epsilon > 0$ and any $t_0 \geq 0$ there exists a $\delta(\epsilon, t_0) > 0$ such that $|x(t, t_0, x_0)| < \epsilon$ for all $t \geq t_0$ whenever $|x_0| < \delta(\epsilon, t_0)$ and $x(t, t_0, x_0) \in B_h(x_e)$ for some $h > 0$. That is, the equilibrium is stable if when the system (5) starts close to x_e, then it will stay close to it. Note that stability is a property of an equilibrium, not a system. A system is stable if all its equilibrium points are stable. Stability in the sense of Lyapunov is a local property. Also, notice that the definition of stability is for a single equilibrium $x_e \in \mathbb{R}^n$ but actually such an equilibrium is a trajectory of points that satisfy the differential equation in (5). That is, the equilibrium x_e is a solution to the differential equation (5), $x(t, t_0, x_0) = x_e$ for $t \geq 0$. We call any set such that when the initial condition of (5) starts in the set and stays in the set for all $t \geq 0$, an *invariant set*. As an example, if $x_e = 0$ is an equilibrium, then the set containing only the point x_e is an invariant set, for (5) [38].

If δ is independent of t_0, that is, if $\delta = \delta(\epsilon)$, then the equilibrium x_e is said to be *uniformly stable*. If in (5) f does not depend on time (i.e., $f(x)$), then x_e being stable is equivalent to it being uniformly stable. Uniform stability is also a local property.

The equilibrium $x_e = 0$ of (5) is said to be *asymptotically stable* if it is stable and for every $t_0 \geq 0$ there exists $\eta(t_0) > 0$ such that $\lim_{t\to\infty} |x(t, t_0, x_0)| = 0$ whenever $|x_0| < \eta(t_0)$. That is, it is asymptotically stable if when it starts close to the equilibrium it will converge to it. Asymptotic stability is also a local property. It is a stronger stability property since it requires that the solutions to the ordinary differential equation converge to zero in addition to what is required for stability in the sense of Lyapunov.

The equilibrium $x_e = 0$ of (5) is said to be *uniformly asymptotically stable* if it is uniformly stable and for every $\epsilon > 0$ and and $t_0 \geq 0$, there exist a $\delta_0 > 0$ independent of t_0 and ϵ, and a $T(\epsilon) > 0$ independent of t_0, such that $|x(t, t_0, x_0) - x_e| \leq \epsilon$ for all

$t \geq t_0 + T(\epsilon)$ whenever $|x_0 - x_e| < \delta(\epsilon)$. Again, if in (5) f does not depend on time (i.e., $f(x)$), then x_e being asymptotically stable is equivalent to it being uniformly asymptotically stable. Uniform asymptotic stability is also a local property.

The set $X_d \subset \mathbb{R}^n$ of all $x_0 \in \mathbb{R}^n$ such that $|x(t, t_0, x_0)| \to 0$ as $t \to \infty$ is called the *domain of attraction* of the equilibrium $x_e = 0$ of (5). The equilibrium $x_e = 0$ is said to be *asymptotically stable* in the large if $X_d \subset \mathbb{R}^n$. That is, an equilibrium is asymptotically stable in the large if no matter where the system starts, its state converges to the equilibrium asymptotically. This is a global property as opposed to the earlier stability definitions that characterized local properties. This means that for asymptotic stability in the large, the local property of asymptotic stability holds for $B_h(x_e)$ with $h = \infty$ (i.e., on the whole state–space).

The equilibrium $x_e = 0$ is said to be *exponentially stable* if there exists an $\alpha > 0$ and for every $\epsilon > 0$ there exists a $\delta(\epsilon) > 0$ such that $|x(t, t_0, x_0)| \leq \epsilon e^{-\alpha(t-t_0)}$, whenever $|x_0| < \delta(\epsilon)$ and $t \geq t_0 \geq 0$. The constant α is sometimes called the *rate of convergence*. Exponential stability is sometimes said to be a 'stronger' form of stability since in its presence we know that system trajectories decrease exponentially to zero. It is a local property; here is its global version. The equilibrium point $x_e = 0$ is *exponentially stable in the large* if there exists $\alpha > 0$ and for any $\beta > 0$ there exists $\epsilon(\beta) > 0$ such that $|x(t, t_0, x_0)| \leq \epsilon(\beta) e^{-\alpha(t-t_0)}$, whenever $|x_0| < \beta$ and $t \geq t_0 \geq 0$.

An equilibrium that is not stable is called *unstable*.

Closely related to stability is the concept of *boundedness*, which is, however, a global property of a system in the sense that it applies to trajectories (solutions) of the system that can be defined over all of the state–space [38].

A solution $x(t, t_0, x_0)$ of (5) is *bounded* if there exists a $\beta > 0$, that may depend on each solution, such that $|x(t, t_0, x_0)| < \beta$ for all $t \geq t_0 \geq 0$. A system is said to possess *Lagrange stability* if for each $t_0 \geq 0$ and $x_0 \in \mathbb{R}^n$, the solution $x(t, t_0, x_0)$ is bounded. If an equilibrium is asymptotically stable in the large or exponentially stable in the large then the system for which the equilibrium is defined is also Lagrange stable (but not necessarily vice versa). Also, if an equilibrium is stable, it does not imply that the system for which the equilibrium is defined is Lagrange stable since there may be a way to pick x_0 such that it is near an unstable equilibrium and $x(t, t_0, x_0) \to \infty$ as $t \to \infty$.

The solutions $x(t, t_0, x_0)$ are *uniformly bounded* if for any $\alpha > 0$ and $t_0 \geq 0$, there exists a $\beta(\alpha) > 0$ (independent of t_0) such that if $|x_0| < \alpha$, then $|x(t, t_0, x_0)| < \beta(\alpha)$ for all $t \geq t_0 \geq 0$. If the solutions are uniformly bounded then they are bounded and the system is Lagrange stable.

The solutions $x(t, t_0, x_0)$ are said to be *uniformly ultimately bounded* if there exists some $B > 0$, and if corresponding to any $\alpha > 0$ and $t_0 > 0$ there exists a $T(\alpha) > 0$ (independent of t_0) such that $|x_0| < \alpha$ implies that $|x(t, t_0, x_0)| < B$ for all $t \geq t_0 + T(\alpha)$. Hence, a system is said to be uniformly ultimately bounded if eventually all trajectories end up in a $B-$neighborhood of the origin.

2.4. Lyapunov's Stability Method

A. M. Lyapunov invented two methods to analyze stability [38]. In his *indirect method* he showed that if we linearize a system about an equilibrium point, certain conclusions about local stability properties can be made (e.g., if the eigenvalues of the linearized system are in the left half plane then the equilibrium is stable but if one is in the right half plane it is unstable).

In his *direct method* the stability results for an equilibrium $x_e = 0$ of (5) depend on the existence of an appropriate *Lyapunov function* $V : D \to \mathbb{R}$ where $D = \mathbb{R}^n$ for global results (e.g., asymptotic stability in the large) and $D = B_h$ for some $h > 0$, for local results (e.g., stability in the sense of Lyapunov or asymptotic stability). If V is continuously differentiable with respect to its arguments then the derivative of V with respect to t along the solutions of (5) is

$$\dot{V}(t,x) = \frac{\partial V}{\partial t} + \frac{\partial V}{\partial x} f(t,x).$$

As an example, suppose that (5) is autonomous, and let $V(x)$ is a quadratic form $V(x) = x^T Px$ where $x \in \mathbb{R}^n$ and $P = P^T$. Then, $\dot{V}(x) = \frac{\partial V}{\partial x} f(t,x) = \dot{x}^T Px + x^T P\dot{x} = 2x^T P\dot{x}$ [38].

Lyapunov's direct method provides for the following ways to test for stability. The first two are strictly for local properties while the last two have local and global versions.

- *Stable*: If $V(t,x)$ is continuously differentiable, positive definite, and $\dot{V}(t,x) \leq 0$, then $x_e = 0$ is stable.

- *Uniformly stable*: If $V(t,x)$ is continuously differentiable, positive definite, decrescent[1], and $V(t,x) \leq 0$, then $x_e = 0$ is uniformly stable.

- *Uniformly asymptotically stable*: If $V(t,x)$ is continuously differentiable, positive definite, and decrescent, with negative definite $\dot{V}(t,x)$, then $x_e = 0$ is uniformly asymptotically stable (uniformly asymptotically stable in the large if all these properties hold globally).

- *Exponentially stable*: If there exists a continuously differentiable $V(t,x)$ and $c, c_1, c_2, c_3 > 0$ such that

$$c_1 |x|^c \leq V(t,x) \leq c_2 |x|^c, \tag{6}$$

$$\dot{V}(t,x) \leq -c_{31} |x|^c, \tag{7}$$

for all $x \in B_h$ and $t \geq 0$, then $x_e = 0$ is exponentially stable. If there exists a continuously differentiable function $V(t,x)$ and Equations (6) and (7) hold for some $c, c_1, c_2, c_3 > 0$ for all $x \in \mathbb{R}^n$ and $t \geq 0$, then $x_e = 0$ is exponentially stable in the large [38].

2.5. Nonlinear and Impulse Dynamics of Complex Plants

In this section we give two examples of nonlinear dynamical systems that are beyond reach of the classical control theory.

2.5.1. Hybrid Dynamical Systems of Variable Structure

Consider a *hybrid dynamical system of variable structure*, given by $n-$dimensional ODE (see [46])

$$\dot{x} = f(t,x), \tag{8}$$

[1] A C^0-function $V(t,x) : \mathbb{R}^+ \times B_h \to \mathbb{R}(V(t,x) : \mathbb{R}^+ \times \mathbb{R}^n \to \mathbb{R})$ is said to be *decrescent* if there exists a strictly increasing function γ defined on $[0,r)$ for some $r > 0$ (defined on $[0,\infty)$) such that $V(t,x) \leq \gamma(|x|)$ for all $t \geq 0$ and $x \in B_h$ for some $h > 0$.

where $x = x(t) \in \mathbb{R}^n$ and $f = f(t,x) : \mathbb{R}^+ \times \mathbb{R}^n \to \mathbb{R}^n$. Let the domain $G \subset \mathbb{R}^+ \times \mathbb{R}^n$, on which the vector–field $f(t,x)$ is defined, be divided into two subdomains, G^+ and G^-, by means of a smooth $(n-1)$–manifold M. In $G^+ \cup M$, let there be given a vector–field $f^+(t,x)$, and in $G^- \cup M$, let there be given a vector–field $f^-(t,x)$. Assume that both $f^+ = f^+(t,x)$ and $f^- = f^-(t,x)$ are continuous in t and smooth in x. For the system (8), let

$$f = \begin{cases} f^+ & \text{when } x \in G^+ \\ f^- & \text{when } x \in G^- \end{cases}.$$

Under these conditions, a solution $x(t)$ of ODE (8) is well–defined while passing through G until the manifold M is reached.

Upon reaching the manifold M, in physical systems with inertia, the transition

$$\text{from} \quad \dot{x} = f^-(t,x) \quad \text{to} \quad \dot{x} = f^+(t,x)$$

does not take place instantly on reaching M, but after some delay. Due to this delay, the solution $x(t)$ oscillates about M, $x(t)$ being displaced along M with some mean velocity.

As the delay tends to zero, the limiting motion and velocity along M are determined by the *linear homotopy ODE*

$$\dot{x} = f^0(t,x) \equiv (1-\alpha)\, f^-(t,x) + \alpha\, f^+(t,x), \tag{9}$$

where $x \in M$ and $\alpha \in [0,1]$ is such that the *linear homotopy segment* $f^0(t,x)$ is tangential to M at the point x, i.e., $f^0(t,x) \in T_xM$, where T_xM is the tangent space to the manifold M at the point x.

The vector–field $f^0(t,x)$ of the system (9) can be constructed as follows: at the point $x \in M$, $f^-(t,x)$ and $f^+(t,x)$ are given and their ends are joined by the linear homotopy segment. The point of intersection between this segment and T_xM is the end of the required vector–field $f^0(t,x)$. The vector function $x(t)$ which satisfies (8) in G^- and G^+, and (9) when $x \in M$, can be considered as a *solution* of (8) in a *general sense*.

However, there are cases in which the solution $x(t)$ cannot consist of a finite or even countable number of arcs, each of which passes through G^- or G^+ satisfying (8), or moves along the manifold M and satisfies the homotopic ODE (9). To cover such cases, assume that the vector–field $f = f(t,x)$ in ODE (8) is a Lebesgue–measurable function in a domain $G \subset \mathbb{R}^+ \times \mathbb{R}^n$, and that for any closed bounded domain $D \subset G$ there exists a summable function $K(t)$ such that almost everywhere in D we have $|f(t,x)| \leq K(t)$. Then the absolutely continuous vector function $x(t)$ is called the *generalized solution* of the ODE (8) *in the sense of Filippov* (see [46]) if for almost all t, the vector $\dot{x} = \dot{x}(t)$ belongs to the least convex closed set containing all the limiting values of the vector field $f(t,x^*)$, where x^* tends towards x in an arbitrary manner, and the values of the function $f(t,x^*)$ on a set of measure zero in $\mathbb{R}^n$ are ignored.

Such *hybrid systems* of variable structure occur in the study of nonlinear electric networks (endowed with electronic switches, relays, diodes, rectifiers, etc.), in models of both natural and artificial neural networks, as well as in feedback control systems (usually with continuous–time plants and digital controllers/filters).

2.5.2. Impulse Dynamics of Kicks and Spikes

The Spike Function. Recall that the *Dirac's δ–function* (also called the *impulse function* in the systems and signals theory) represents a *limit* of the *Gaussian bell–shaped curve*

$$g(t,\alpha) = \frac{1}{\sqrt{\pi\alpha}} \mathrm{e}^{-t^2/\alpha} \quad \text{(with parameter } \alpha \to 0) \tag{10}$$

() where the factor $1/\sqrt{\pi\alpha}$ serves for the *normalization* of (10),

$$\int_{-\infty}^{+\infty} \frac{dt}{\sqrt{\pi\alpha}} \mathrm{e}^{-t^2/\alpha} = 1, \tag{11}$$

i.e., the *area* under the *pulse* is equal to *unity*. In (10), the smaller α the higher the peak. In other words,

$$\delta(t) = \lim_{\alpha \to 0} \frac{1}{\sqrt{\pi\alpha}} \mathrm{e}^{-t^2/\alpha}, \tag{12}$$

which is a pulse so short that outside of $t = 0$ it vanishes, whereas at $t = 0$ it still remains normalized according to (11). Therefore, we get the usual definition of the δ–function:

$$\begin{aligned} \delta(t) &= 0 \quad \text{for} \quad \mathrm{t} \neq 0, \\ \int_{-\epsilon}^{+\epsilon} \delta(t)\, dt &= 1, \end{aligned} \tag{13}$$

where ϵ may be arbitrarily small. Instead of centering the δ–pulse around $t = 0$, we can center it around any other time t_0 so that (13) is transformed into

$$\begin{aligned} \delta(t - t_0) &= 0 \quad \text{for} \quad \mathrm{t} \neq t_0, \\ \int_{t_0-\epsilon}^{t_0+\epsilon} \delta(t - t_0)\, dt &= 1. \end{aligned} \tag{14}$$

Another well–known fact is that the integral of the δ–function is the *Heaviside's step function*

$$H(T) = \int_{-\infty}^{T} \delta(t)\, dt = \begin{cases} 0 & \text{for} \quad T < 0 \\ 1 & \text{for} \quad T > 0 \\ (\frac{1}{2} & \text{for} \quad T = 0) \end{cases} . \tag{15}$$

Now we can perform several generalizations of the relation (15). First, we have

$$\int_{-\infty}^{T} \delta(ct - t_0)\, dt = \begin{cases} 0 & \text{for} \quad T < t_0/c \\ 1/c & \text{for} \quad T > t_0/c \\ \frac{1}{2c} & \text{for} \quad T = t_0/c \end{cases} .$$

More generally, we can introduce the so–called *phase function* $\phi(t)$, (e.g., $\phi(t) = ct - t_0$) which is continuous at $t = t_0$ but its time derivative $\dot{\phi}(t) \equiv \frac{d\phi(t)}{dt}$ is discontinuous at $t = t_0$ (yet positive, $\dot{\phi}(t) > 0$), and such that

$$\int_{-\infty}^{T} \delta(\phi(t))\, dt = \begin{cases} 0 & \text{for} \quad T < t_0 \\ 1/\dot{\phi}(t_0) & \text{for} \quad T > t_0 \\ \frac{1}{\dot{\phi}(t_0)} & \text{for} \quad T = t_0 \end{cases} .$$

Finally, we come the the *spike function* $\delta(\phi(t))\dot{\phi}(t)$, which like $\delta-$function represents a *spike* at $t = t_0$, such that the normalization criterion (14) is still valid,

$$\int_{t_0-\epsilon}^{t_0+\epsilon} \delta(\phi(t))\dot{\phi}(t)\,dt = 1.$$

Deterministic Delayed Kicks. Following Haken [47], we consider the mechanical example of a soccer ball that is kicked by a soccer player and rolls over grass, whereby its motion will be slowed down. In our opinion, this is a perfect model for all 'shooting–like' actions of the human operator.

We start with the Newton's (second) law of motion, $m\dot{v} = force$, and in order to get rid of superfluous constants, we put temporarily $m = 1$. The $force$ on the r.h.s. consists of the damping force $-\gamma v(t)$ of the grass (where γ is the damping constant) and the sharp force $F(t) = s\delta(t-\sigma)$ of the individual kick occurring at time $t = \sigma$ (where s is the strength of the kick, and δ is the Dirac's 'delta' function). In this way, the single–kick equation of the ball motion becomes

$$\dot{v} = -\gamma v(t) + s\delta(t-\sigma), \tag{16}$$

with the general solution

$$v(t) = sG(t-\sigma),$$

where $G(t-\sigma)$ is the Green's function[2]

$$G(t-\sigma) = \begin{cases} 0 & \text{for } t < \sigma \\ \mathrm{e}^{-\gamma(t-\sigma)} & \text{for } t \geq \sigma \end{cases}.$$

Now, we can generalize the above to N kicks with individual strengths s_j, occurring at a sequence of times $\{\sigma_j\}$, so that the total kicking force becomes

$$F(t) = \sum_{j=1}^{N} s_j\delta(t-\sigma_j).$$

In this way, we get the multi–kick equation of the ball motion

$$\dot{v} = -\gamma v(t) + \sum_{j=1}^{N} s_j\delta(t-\sigma_j),$$

with the general solution

$$v(t) = \sum_{j=1}^{N} s_j G(t-\sigma_j). \tag{17}$$

[2]This is the Green's function of the first order system (16). Similarly, the Green's function

$$G(t-\sigma) = \begin{cases} 0 & \text{for } t < \sigma \\ (t-\sigma)\mathrm{e}^{-\gamma(t-\sigma)} & \text{for } t \geq \sigma \end{cases}$$

corresponds to the second order system

$$\left(\frac{d}{dt} + \gamma\right)^2 G(t-\sigma) = \delta(t-\sigma).$$

As a final generalization, we would imagine that the kicks are continuously exerted on the ball, so that kicking force becomes

$$F(t) = \int_{t_0}^{T} s(\sigma)\delta(t-\sigma)d\sigma \equiv \int_{t_0}^{T} d\sigma F(\sigma)\delta(t-\sigma),$$

so that the continuous multi–kick equation of the ball motion becomes

$$\dot{v} = -\gamma v(t) + \int_{t_0}^{T} s(\sigma)\delta(t-\sigma)d\sigma \equiv -\gamma v(t) + \int_{t_0}^{T} d\sigma F(\sigma)\delta(t-\sigma),$$

with the general solution

$$v(t) = \int_{t_0}^{T} d\sigma F(\sigma)G(t-\sigma) = \int_{t_0}^{T} d\sigma F(\sigma)\mathrm{e}^{-\gamma(t-\sigma)}. \tag{18}$$

Random Kicks and Langevin Equations. We now denote the times at which kicks occur by t_j and indicate their direction in a one–dimensional game by $(\pm 1)_j$, where the choice of the plus or minus sign is random (e.g., throwing a coin). Thus the kicking force can be written in the form

$$F(t) = s\sum_{j=1}^{N} \delta(t-t_j)(\pm 1)_j, \tag{19}$$

where for simplicity we assume that all kicks have the same strength s. When we observe many games, then we may perform an average $< ... >$ over all these different *performances*,

$$< F(t) >= s < \sum_{j=1}^{N} \delta(t-t_j)(\pm 1)_j > . \tag{20}$$

Since the direction of the kicks is assumed to be independent of the time at which the kicks happen, we may split (20) into the product

$$< F(t) >= s < \sum_{j=1}^{N} \delta(t-t_j) >< (\pm 1)_j > .$$

As the kicks are assumed to happen with equal frequency in both directions, we get the cancellation

$$< (\pm 1)_j >= 0,$$

which implies that the average kicking force also vanishes,

$$< F(t) >= 0.$$

In order to characterize the strength of the force (19), we consider a quadratic expression in F, e.g., by calculating the *correlation function* for two times t, t',

$$< F(t)F(t') >= s^2 < \sum_{j} \delta(t-t_j)(\pm 1)_j \sum_{k} \delta(t'-t_k)(\pm 1)_k > .$$

As the ones for $j \neq k$ will cancel each other and for $j = k$ will become 1, the correlation function becomes a single sum

$$< F(t)F(t') >= s^2 < \sum_j \delta(t - t_j)\delta(t' - t_k) >, \tag{21}$$

which is usually evaluated by assuming the *Poisson process* for the times of the kicks.

Now, proper description of random motion is given by *Langevin rate equation*, which describes the *Brownian motion*: when a particle is immersed in a fluid, the velocity of this particle is slowed down by a force proportional to its velocity and the particle undergoes a zig–zag motion (the particle is steadily pushed by much smaller particles of the liquid in a random way). In physical terminology, we deal with the behavior of a system (particle) which is coupled to a *heat bath* or reservoir (namely the liquid). The heat bath has two effects:

1. It decelerates the mean motion of the particle; and
2. It causes statistical fluctuation.

The standard Langevin equation has the form

$$\dot{v} = -\gamma v(t) + F(t), \tag{22}$$

where $F(t)$ is a *fluctuating force* with the following properties:

1. Its statistical average (20) vanishes; and
2. Its correlation function (21) is given by

$$< F(t)F(t') >= Q\delta(t - t_0), \tag{23}$$

where $t_0 = T/N$ denotes the mean free time between kicks, and $Q = s^2/t_0$ is the *random fluctuation.*

The general solution of the Langevin equation (22) is given by (18).

The average velocity vanishes, $< v(t) >= 0$, as both directions are possible and cancel each other. Using the integral solution (18) we get

$$< v(t)v(t') >=< \int_{t_0}^{t} d\sigma \int_{t_0}^{t'} d\sigma' F(\sigma)F(\sigma') \mathrm{e}^{-\gamma(t-\sigma)} \mathrm{e}^{-\gamma(t'-\sigma')} >,$$

which, in the steady–state, reduces to

$$< v(t)v(t') >= \frac{Q}{2\gamma} \mathrm{e}^{-\gamma(t-\sigma)},$$

and for equal times

$$< v(t)^2 >= \frac{Q}{2\gamma}.$$

If we now repeat all the steps performed so far with $m \neq 1$, the final result reads

$$< v(t)^2 >= \frac{Q}{2\gamma m}. \tag{24}$$

Now, according to thermodynamics, the *mean kinetic energy* of a particle is given by

$$\frac{m}{2} < v(t)^2 >= \frac{1}{2} k_B T, \tag{25}$$

where T is the (absolute) temperature, and k_B is the Boltzman's constant. Comparing (24) and (25), we obtain the important Einstein's result

$$Q = 2\gamma k_B T,$$

which says that whenever there is damping, i.e., $\gamma \neq 0$, then there are random fluctuations (or noise) Q. In other words, fluctuations or noise are inevitable in any physical system. For example, in a resistor (with the resistance R) the electric field E fluctuates with a correlation function (similar to (23))

$$< E(t)E(t') >= 2Rk_B T\delta(t - t_0).$$

This is the simplest example of the so–called *dissipation–fluctuation theorem.*

3. Nonlinear Control Modeling of the Human Operator

In this section we present the basics of modern nonlinear control, as a powerful tool for controlling nonlinear dynamical systems.

3.1. Graphical Techniques for Nonlinear Systems

Graphical techniques preceded modern geometrical techniques in nonlinear control theory. They started with simple plotting tools, like the so–called 'tracer plot'. It is a useful visualization tool for analysis of second order dynamical systems, which just adds time dimension to the standard 2D phase portrait. For example, consider the damped spring governed by

$$\ddot{x} = -k\dot{x} - x, \; x(0) = 1.$$

Its tracer plot is given in Figure 1. Note the stable asymptote reached as $t \to \infty$.

The most important graphical technique is the so–called describing function analysis.

3.1.1. Describing Function Analysis

Describing function analysis extends classical linear control technique, frequency response analysis, for nonlinear systems [37]. It is an *approximate* graphical method mainly used to predict *limit cycles* in nonlinear ODEs.

For example, if we want to predict the existence of limit cycles in the classical *Van der Pol's oscillator* given by

$$\ddot{x} + \alpha(x^2 - 1)\,\dot{x} + x = 0, \tag{26}$$

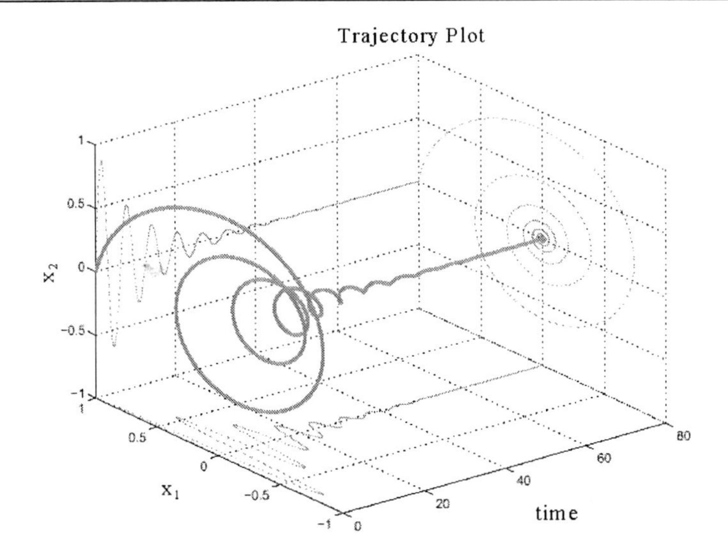

Figure 1. Tracer plot of the damped spring.

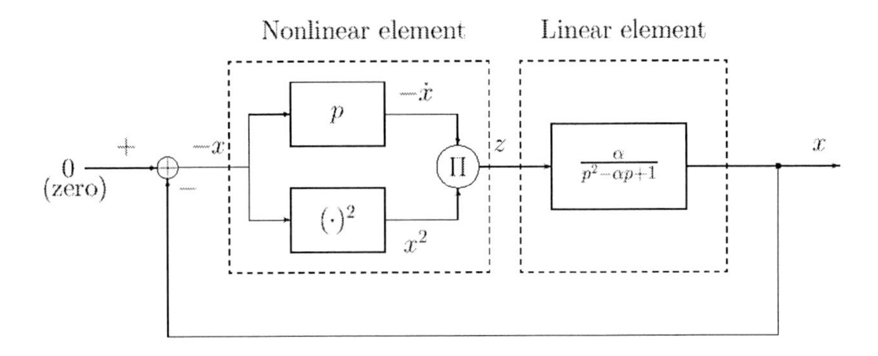

Figure 2. Feedback interpretation of the Van Der Pol oscillator (after [37]). Here p is a (linear) differentiator, and Π a (nonlinear) multiplicator.

we need to rewrite (26) as a *linear* unstable low–pass block and a *nonlinear* block (see Figure 2). In this way, using the nonlinear block substitution, $w := -\dot{x}x^2$, we get

$$\begin{aligned} \ddot{x} - \alpha\dot{x} + x &= \alpha w, \qquad \text{or} \\ x(p^2 - \alpha p + 1) &= \alpha w, \end{aligned}$$

or just considering the transfer function from w to x,

$$\frac{x}{w} = \frac{\alpha}{p^2 - \alpha p + 1}.$$

Now, if we assume that the Van der Pol oscillator does have a limit cycle with a frequency of

$$x(t) = A\sin(wt),$$

so $\dot{x} = Aw\cos(wt)$, therefore the *output* of the nonlinear block is

$$\begin{aligned} z &= -x^2\dot{x} = -A^2\sin^2(wt)\,Aw\cos(wt) \\ &= -\frac{A^3w}{2}(1-\cos(2wt))\cos(wt) \\ &= -\frac{A^3w}{4}(\cos(wt)-\cos(3wt)). \end{aligned}$$

Note how z contains a third harmonic term, but this is attenuated by the low–pass nature of the linear block, and so does not effect the signal in the feedback. So we can approximate z by

$$z \approx \frac{A^3}{4}w\cos(wt) = \frac{A^2}{4}\frac{d}{dt}(-A\sin(wt)).$$

Therefore, the output of the nonlinear block can be approximated by the quasi–linear transfer function which depends on the signal amplitude, A, as well as frequency. The frequency response function of the quasi–linear element is obtained by substituting $p \equiv s = iw$,

$$N(A,w) = \frac{A^2}{4}(iw).$$

Since the system is assumed to contain a sinusoidal oscillation,

$$\begin{aligned} x &= A\sin(wt) = G(iw)\,z \\ &= G(iw)\,N(A,w)\,(-x), \end{aligned}$$

where $G(iw)$ is the transfer function of the linear block. This implies that,

$$\frac{x}{-x} = -1 = G(iw)\,N(A,w),$$

so

$$1 + \frac{A^2(iw)}{4}\frac{\alpha}{(iw)^2 - \alpha(iw) + 1} = 0,$$

which solving gives,

$$A = 2, \qquad \omega = 1,$$

which is independent of α. Note that in terms of the Laplace variable $p \equiv s$, the closed loop characteristic equation of the system is

$$1 + \frac{A^2(iw)}{4}\frac{\alpha}{p^2 - \alpha p + 1} = 0,$$

whose eigenvalues are

$$\lambda_{1,2} = -\frac{1}{8}\alpha(A^2-4) \pm \sqrt{\frac{\alpha^2(A^2-4)^2}{64}} - 1.$$

Corresponding to $A = 2$ gives eigenvalues of $\lambda_{1,2} = \pm i$ indicating an existence of a limit cycle of amplitude 2 and frequency 1 (see Figure 3). If $A > 2$ eigenvalues are negative real, so stable, and the same holds for $A < 2$. The approximation of the nonlinear block with $(A2/4)(iw)$ is called the *describing function*. This technique is useful because most limit cycles are approximately sinusoidal and most linear elements are low–pass in nature. So most of the higher harmonics, if they existed, are attenuated and lost.

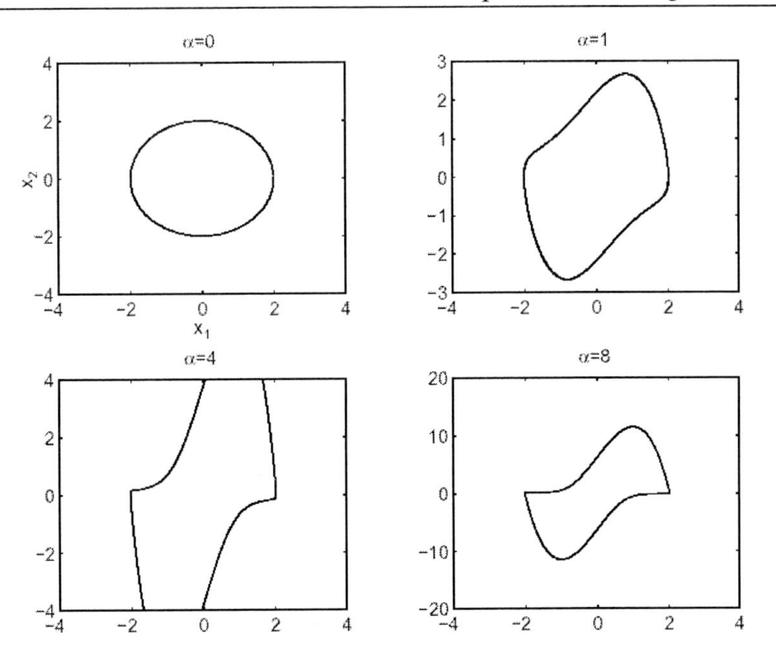

Figure 3. Limit cycle of the Van der Pol oscillator (after [37]). Approximation is reasonable for small α, but error, (amplitude should be equal to 2), grows as α is increased.

3.2. Feedback Linearization

The idea of feedback linearization is to algebraically transform the nonlinear system dynamics into a fully or partly linear one so that the linear control techniques can be applied. Note that this is not the same as a conventional linearization using Jacobians. In this subsection we will present the modern, geometrical, Lie–derivative based techniques for exact feedback linearization of nonlinear control systems.

3.2.1. The Lie Derivative and Lie Bracket in Control Theory

Recall that given a scalar function $h(x)$ and a vector–field $f(x)$, we define a new scalar function, $\mathcal{L}_f h := \nabla h f$, which is the *Lie derivative* of h w.r.t. f, i.e., the directional derivative of h along the direction of the vector f (see [4, 5]). Repeated Lie derivatives can be defined recursively:

$$\begin{aligned} \mathcal{L}_f^0 h &= h, \\ \mathcal{L}_f^i h &= \mathcal{L}_f\left(\mathcal{L}_f^{i-1} h\right) = \nabla\left(\mathcal{L}_f^{i-1} h\right) f, \qquad \text{for } i = 1, 2, ... \end{aligned}$$

Or given another vector–field, g, then $\mathcal{L}_g \mathcal{L}_f h(x)$ is defined as

$$\mathcal{L}_g \mathcal{L}_f h = \nabla\left(\mathcal{L}_f h\right) g.$$

For example, if we have a control system

$$\begin{aligned} \dot{x} &= f(x), \\ y &= h(x), \end{aligned}$$

with the state $x = x(t)$ and the the output y, then the derivatives of the output are:

$$\begin{aligned} \dot{y} &= \frac{\partial h}{\partial x}\dot{x} = \mathcal{L}_f h, \quad \text{and} \\ \ddot{y} &= \frac{\partial L_f h}{\partial x}\dot{x} = \mathcal{L}_f^2 h. \end{aligned}$$

Also, recall that the curvature of two vector–fields, g_1, g_2, gives a non–zero Lie bracket, $[g_1, g_2]$ (see Figure 4). Lie bracket motions can generate new directions in which the system can move.

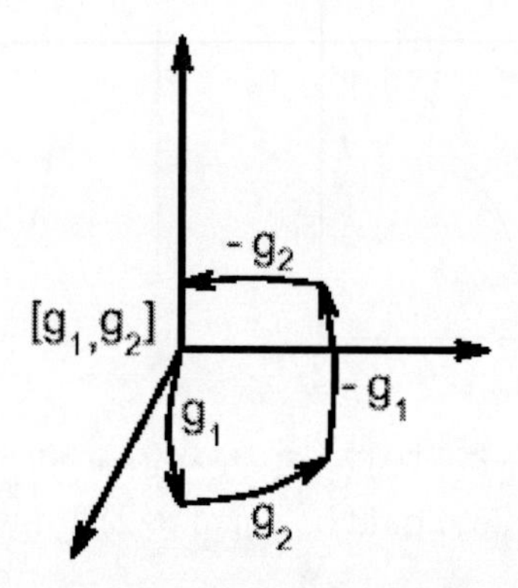

Figure 4. 'Lie bracket motion' is possible by appropriately modulating the control inputs (from [42]).

In general, the Lie bracket of two vector–fields, $f(x)$ and $g(x)$, is defined by

$$[f, g] := ad_f g := \nabla g f - \nabla f g := \frac{\partial g}{\partial x} f - \frac{\partial f}{\partial x} g,$$

where $\nabla f := \partial f / \partial x$ is the Jacobian matrix. We can define Lie brackets recursively,

$$\begin{aligned} ad_f^0 g &= g, \\ ad_f^i g &= [f, ad_f^{i-1} g], \qquad \text{for } i = 1, 2, ... \end{aligned}$$

Lie brackets have the properties of bilinearity, skew–commutativity and Jacobi identity.

For example, if

$$f = \begin{pmatrix} \cos x_2 \\ x_1 \end{pmatrix}, \qquad g = \begin{pmatrix} x_1 \\ 1 \end{pmatrix},$$

then we have

$$\begin{aligned} [f, g] &= \begin{pmatrix} 1 & 0 \\ 0 & 0 \end{pmatrix} \begin{pmatrix} \cos x_2 \\ x_1 \end{pmatrix} - \begin{pmatrix} 0 & -\sin x_2 \\ 1 & 0 \end{pmatrix} \begin{pmatrix} x_1 \\ 1 \end{pmatrix} \\ &= \begin{pmatrix} \cos x_2 + \sin x_2 \\ -x_1 \end{pmatrix}. \end{aligned}$$

3.2.2. Input/Output Linearization

Given the single–input single–output (SISO) system

$$\begin{aligned} \dot{x} &= f(x) + g(x)\,u, \\ y &= h(x), \end{aligned} \tag{27}$$

we want to formulate a linear differential equation relation between output y and a new input v. We will investigate (see [41, 43, 37]):

- How to generate a linear input/output relation.
- What are the internal dynamics and zero–dynamics associated with the input/output linearization?
- How to design stable controllers based on the I/O linearization.

This linearization method will be exact in a finite domain, rather than tangent as in the local linearization methods, which use Taylor series approximation. Nonlinear controller design using the technique is called exact feedback linearization.

3.2.3. Algorithm for Exact Feedback Linearization

We want to find a nonlinear compensator such that the closed–loop system is linear (see Figure 5). We will consider only affine SISO systems of the type (27), i.e, $\dot{x} = f(x) + g(x)\,u$, $y = h(x)$, and we will try to construct a *control law* of the form

$$u = p(x) + q(x)\,v, \tag{28}$$

where v is the setpoint, such that the *closed–loop system*

$$\begin{aligned} \dot{x} &= f(x) + g(x)\,p(x) + g(x)\,q(x)\,v, \\ y &= h(x), \end{aligned}$$

is linear from command v to y.

The main idea behind the feedback linearization construction is to find a nonlinear change of coordinates which transforms the original system into one which is linear and controllable, in particular, a chain of integrators. The difficulty is finding the output function $h(x)$ which makes this construction possible.

We want to design an exact nonlinear feedback controller. Given the nonlinear affine system, $\dot{x} = f(x) + g(x)$, $y = h(x)$, we want to find the controller functions $p(x)$ and $q(x)$. The unknown functions inside our controller (28) are given by:

$$\begin{aligned} p(x) &= \frac{-\left(\mathcal{L}_f^r h(x) + \beta_1 \mathcal{L}_f^{r-1} h(x) + \ldots + \beta_{r-1} \mathcal{L}_f h(x) + \beta_r h(x)\right)}{\mathcal{L}_g \mathcal{L}_f^{r-1} h(x)}, \\ q(x) &= \frac{1}{\mathcal{L}_g \mathcal{L}_f^{r-1} h(x)}, \end{aligned} \tag{29}$$

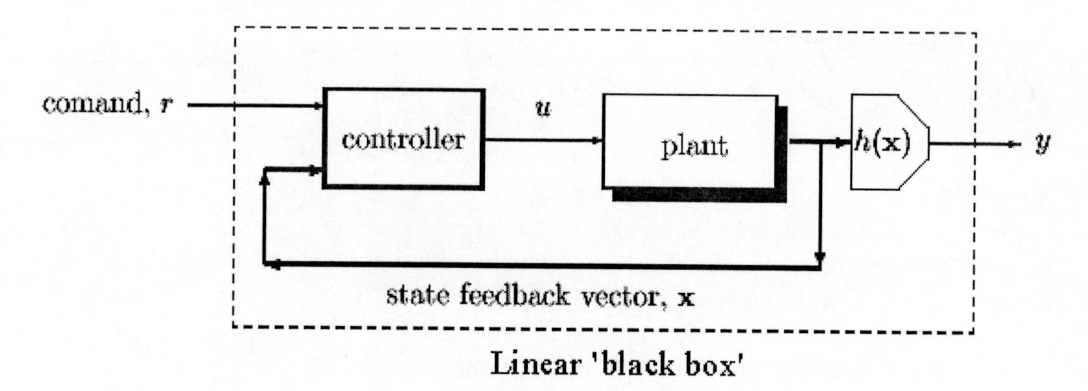

Figure 5. Feedback linearization.

which are comprised of Lie derivatives, $\mathcal{L}_f h(x)$. Here, the *relative order*, r, is the smallest integer r such that $\mathcal{L}_g \mathcal{L}_f^{r-1} h(x) \neq 0$. For linear systems r is the difference between the number of poles and zeros.

To obtain the *desired response*, we choose the r parameters in the β polynomial to describe how the output will respond to the setpoint, v (pole–placement).

$$\frac{d^r y}{dt^r} + \beta_1 \frac{d^{r-1} y}{dt^{r-1}} + ... + \beta_{r-1} \frac{dy}{dt} + \beta_r y = v.$$

Here is the proposed algorithm [41, 43, 37]):

1. Given nonlinear SISO process, $\dot{x} = f(x, u)$, and output equation $y = h(x)$, then:
2. Calculate the relative order, r.
3. Choose an rth order desired linear response using pole–placement technique (i.e., select β). For this could be used a simple rth order low–pass filter such as a Butterworth filter.
4. Construct the exact linearized nonlinear controller (29), using Lie derivatives and perhaps a symbolic manipulator (Mathematica or Maple).
5. Close the loop and obtain a linear input–output black–box (see Figure 5).
6. Verify that the result is actually linear by comparing with the desired response.

3.3. Controllability

3.3.1. Linear Controllability

A system is *controllable* if the set of all states it can reach from initial state $x_0 = x(0)$ at the fixed time $t = T$ contains a ball $\mathcal{B}$ around x_0. Again, a system is *small time locally controllable* (STLC) iff the ball $\mathcal{B}$ for $t \leq T$ contains a neighborhood of x_0.[3]

[3]The above definition of controllability tells us only whether or not something can reach an open neighborhood of its starting point, but does not tell us how to do it. That is the point of the *trajectory generation*.

In the case of a linear system in the standard state–space form

$$\dot{x} = Ax + Bu, \tag{30}$$

where A is the $n \times n$ *state matrix* and B is the $m \times n$ *input matrix*, all controllability definitions coincide, i.e.,

$$\begin{aligned} 0 &\rightarrow x(T), \\ x(0) &\rightarrow 0, \\ x(0) &\rightarrow x(T), \end{aligned}$$

where T is either fixed or free.

Rank condition states: System (30) is controllable iff the matrix

$$W_n = \left(B\, AB \ldots A^{n-1}B\right) \qquad \text{has full rank.}$$

In the case of nonlinear systems the corresponding result is obtained using the formalism of Lie brackets, as Lie algebra is to nonlinear systems as matrix algebra is to linear systems.

3.3.2. Nonlinear Controllability

Nonlinear MIMO–systems are generally described by differential equations of the form (see [41, 44, 42]):

$$\dot{x} = f(x) + g_i(x)\, u^i, \qquad (i = 1, \ldots, n), \tag{31}$$

defined on a smooth $n-$manifold M, where $x \in M$ represents the state of the control system, $f(x)$ and $g_i(x)$ are vector–fields on M and the u^i are control inputs, which belong to a set of *admissible controls*, $u^i \in U$. The system (31) is called *driftless*, or *kinematic*, or *control linear* if $f(x)$ is identically zero; otherwise, it is called a *system with drift*, and the vector–field $f(x)$ is called the *drift term*. The flow $\phi_t^g(x_0)$ represents the solution of the differential equation $\dot{x} = g(x)$ at time t starting from x_0. Geometrical way to understand the *controllability* of the system (31) is to understand the geometry of the vector–fields $f(x)$ and $g_i(x)$.

Example: Car–Parking Using Lie Brackets. In this popular example, the driver has two different transformations at his disposal. He can turn the steering wheel, or he can drive the car forward or back. Here, we specify the state of a car by four coordinates: the (x, y) coordinates of the center of the rear axle, the direction θ of the car, and the angle ϕ between the front wheels and the direction of the car. L is the constant length of the car. Therefore, the configuration manifold of the car is 4D, $M := (x, y, \theta, \phi)$.

Using (31), the driftless car kinematics can be defined as:

$$\dot{x} = g_1(x)\, u_1 + g_2(x)\, u_2, \tag{32}$$

with two vector–fields $g_1, g_2 \in \mathcal{X}^k(M)$.

The infinitesimal transformations will be the vector–fields

$$g_1(x) \equiv \text{DRIVE} = \cos\theta\frac{\partial}{\partial x} + \sin\theta\frac{\partial}{\partial y} + \frac{\tan\phi}{L}\frac{\partial}{\partial\theta} \equiv \begin{pmatrix} \cos\theta \\ \sin\theta \\ \frac{1}{L}\tan\phi \\ 0 \end{pmatrix},$$

and

$$g_2(x) \equiv \text{STEER} = \frac{\partial}{\partial\phi} \equiv \begin{pmatrix} 0 \\ 0 \\ 0 \\ 1 \end{pmatrix}.$$

Now, STEER and DRIVE do not commute; otherwise we could do all your steering at home before driving of on a trip. Therefore, we have a Lie bracket

$$[g_2, g_1] \equiv [\text{STEER}, \text{DRIVE}] = \frac{1}{L\cos^2\phi}\frac{\partial}{\partial\theta} \equiv \text{ROTATE}.$$

The operation $[g_2, g_1] \equiv \text{ROTATE} \equiv [\text{STEER,DRIVE}]$ is the infinitesimal version of the sequence of transformations: steer, drive, steer back, and drive back, i.e.,

$$\{\text{STEER}, \text{DRIVE}, \text{STEER}^{-1}, \text{DRIVE}^{-1}\}.$$

Now, ROTATE can get us out of some parking spaces, but not tight ones: we may not have enough room to ROTATE out. The usual tight parking space restricts the DRIVE transformation, but not STEER. A truly tight parking space restricts STEER as well by putting your front wheels against the curb.

Fortunately, there is still another commutator available:

$$\begin{aligned} [g_1, [g_2, g_1]] &\equiv [\text{DRIVE}, [\text{STEER}, \text{DRIVE}]] = [[g_1, g_2], g_1] \equiv \\ [\text{DRIVE}, \text{ROTATE}] &= \frac{1}{L\cos^2\phi}\left(\sin\theta\frac{\partial}{\partial x} - \cos\theta\frac{\partial}{\partial y}\right) \equiv \text{SLIDE}. \end{aligned}$$

The operation $[[g_1, g_2], g_1] \equiv \text{SLIDE} \equiv [\text{DRIVE,ROTATE}]$ is a displacement at right angles to the car, and can get us out of any parking place. We just need to remember to steer, drive, steer back, drive some more, steer, drive back, steer back, and drive back:

$$\{\text{STEER}, \text{DRIVE}, \text{STEER}^{-1}, \text{DRIVE}, \text{STEER}, \text{DRIVE}^{-1}, \text{STEER}^{-1}, \text{DRIVE}^{-1}\}.$$

We have to reverse steer in the middle of the parking place. This is not intuitive, and no doubt is part of the problem with parallel parking.

Thus from only two controls u_1 and u_2 we can form the vector fields DRIVE $\equiv g_1$, STEER $\equiv g_2$, ROTATE $\equiv [g_2, g_1]$, and SLIDE $\equiv [[g_1, g_2], g_1]$, allowing us to move anywhere in the configuration manifold M. The car kinematics $\dot{x} = g_1 u_1 + g_2 u_2$ is thus expanded as:

$$\begin{pmatrix} \dot{x} \\ \dot{y} \\ \dot{\theta} \\ \dot{\phi} \end{pmatrix} = \text{DRIVE}\cdot u_1 + \text{STEER}\cdot u_2 \equiv \begin{pmatrix} \cos\theta \\ \sin\theta \\ \frac{1}{L}\tan\phi \\ 0 \end{pmatrix}\cdot u_1 + \begin{pmatrix} 0 \\ 0 \\ 0 \\ 1 \end{pmatrix}\cdot u_2\,.$$

The *parking theorem* says: One can get out of any parking lot that is larger than the car.

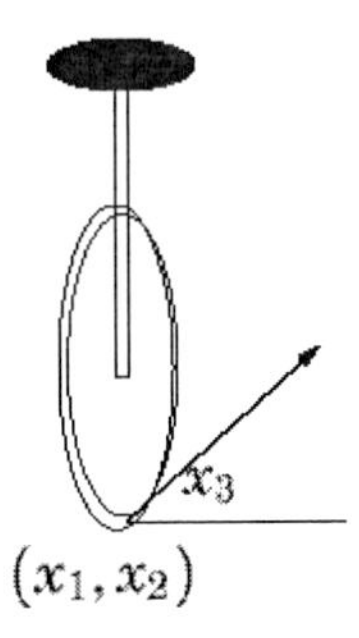

Figure 6. The unicycle.

The Unicycle Example. Now, consider the unicycle example (see Figure 6). Here we have

$$g_1 = \begin{pmatrix} \cos x_3 \\ \sin x_3 \\ 0 \end{pmatrix}, \qquad g_2 = \begin{pmatrix} 0 \\ 0 \\ 1 \end{pmatrix},$$

$$[g_1, g_2] = \begin{pmatrix} \sin x_3 \\ -\cos x_3 \\ 0 \end{pmatrix}.$$

The unicycle system is full rank and therefore controllable.

3.3.3. Controllability Condition

Nonlinear controllability is an extension of linear controllability. The nonlinear SIMO system

$$\dot{x} = f(x) + g(x)\,u$$

is controllable if the set of vector–fields

$$\{g, [f, g], ..., [f^{n-1}, g]\}$$

is independent.

For example, for the kinematic kar system of the form (32), the *nonlinear controllability criterion* reads: If the Lie bracket tree:
$g_1,\ g_2,\ [g_1, g_2],\ [[g_1, g_2], g_1],\ [[g_1, g_2], g_2],\ [[[g_1, g_2], g_1], g_1],\ [[[g_1, g_2], g_1], g_2],$ $[[[g_1, g_2], g_2], g_1],\ [[[g_1, g_2], g_2], g_2], \ldots$
– has *full rank* then the system is *controllable* [41, 44, 42]. In this case the combined input

$$(u_1, u_2) = \begin{cases} (1, 0), & t \in [0, \varepsilon] \\ (0, 1), & t \in [\varepsilon, 2\varepsilon] \\ (-1, 0), & t \in [2\varepsilon, 3\varepsilon] \\ (0, -1), & t \in [3\varepsilon, 4\varepsilon] \end{cases}$$

gives the motion $x(4\varepsilon) = x(0) + \varepsilon^2\,[g_1, g_2] + O(\varepsilon^3)$, with the flow given by

$$F_t^{[g_1, g_2]} = \lim_{n\to\infty} \left(F^{-g_2}_{\sqrt{t/n}} F^{-g_1}_{\sqrt{t/n}} F^{g_2}_{\sqrt{t/n}} F^{g_1}_{\sqrt{t/n}} \right)^n.$$

3.4. Adaptive Lie–Derivative Control

In this subsection we develop the concept of *machine learning* in the framework of Lie–derivative control formalism (see (3.2.1.) above). Consider an $n-$dimensional, SISO system in the standard affine form (27), rewritten here for convenience:

$$\dot{x}(t) = f(x) + g(x)\,u(t), \qquad y(t) = h(x), \tag{33}$$

As already stated, the feedback control law for the system (33) can be defined using Lie derivatives $\mathcal{L}_f h$ and $\mathcal{L}_g h$ of the system's output h along the vector–fields f and g.

If the SISO system (33) is a relatively simple (quasilinear) system with *relative degree*[4] $= 1$, it can be rewritten in a quasilinear form

$$\dot{x}(t) = \gamma_i(t)\,f_i(x) + d_j(t)\,g_j(x)\,u(t), \tag{34}$$

where γ_i $(i = 1, ..., n)$ and d_j $(j = 1, ..., m)$ are system's parameters, while f_i and g_j are smooth vector–fields.

In this case the feedback control law for *tracking* the *reference signal* $y_R = y_R(t)$ is defined as (see [41, 44])

$$u = \frac{-\mathcal{L}_f h + \dot{y}_R + \alpha\,(y_R - y)}{\mathcal{L}_g h}, \tag{35}$$

where α denotes the feedback gain.

Obviously, the problem of reference signal tracking is relatively simple and straightforward if we know all the system's parameters $\gamma_i(t)$ and $d_j(t)$ of (34). The question is can we apply a similar control law if the system parameters are unknown?

Now we have much harder problem of *adaptive signal tracking*. However, it appears that the feedback control law can be actually cast in a similar form (see [43],[45]):

$$\widehat{u} = \frac{-\widehat{\mathcal{L}_f h} + \dot{y}_R + \alpha\,(y_R - y)}{\widehat{\mathcal{L}_g h}}, \tag{36}$$

where Lie derivatives $\mathcal{L}_f h$ and $\mathcal{L}_g h$ of (35) have been replaced by their estimates $\widehat{\mathcal{L}_f h}$ and $\widehat{\mathcal{L}_g h}$, defined respectively as

$$\widehat{\mathcal{L}_f h} = \widehat{\gamma_i}(t)\,\mathcal{L}_{f_i} h, \qquad \widehat{\mathcal{L}_g h} = \widehat{d_j}(t)\,\mathcal{L}_{g_i} h,$$

in which $\widehat{\gamma_i}(t)$ and $\widehat{d_j}(t)$ are the estimates for $\gamma_i(t)$ and $d_j(t)$.

Therefore, we have the straightforward control law even in the uncertain case, provided that we are able to estimate the unknown system parameters. Probably the best known *parameter update law* is based on the so–called *Lyapunov criterion* (see [43]) and given by

$$\dot{\psi} = -\gamma\,\epsilon\,W, \tag{37}$$

[4]Relative degree equals the number of differentiations of the output function y required to have the input u appear explicitly. Technically, the system (33) is said to have relative degree r at the point x^0 if (see [41, 44])

(i) $\mathcal{L}_g \mathcal{L}_f^k h(x) = 0$ for all x in a neighborhood of x^0 and all $k < r - 1$, and

(ii) $\mathcal{L}_g \mathcal{L}_f^{r-1} h(x^0) \neq 0$,

where $\mathcal{L}_f^k h$ denotes the kth Lie derivative of h along f.

where $\psi = \{\gamma_i - \widehat{\gamma_i}, d_j - \widehat{d_j}\}$ is the parameter estimation error, $\epsilon = y - y_R$ is the output error, and γ is a positive constant, while the matrix W is defined as:

$$
\begin{aligned}
W &= \left[W_1^T \, W_2^T\right]^T, \qquad \text{with} \\
W_1 &= \begin{bmatrix} \mathcal{L}_{f_1} h \\ \vdots \\ \mathcal{L}_{f_n} h \end{bmatrix}, \qquad W_2 = \begin{bmatrix} \mathcal{L}_{g_1} h \\ \vdots \\ \mathcal{L}_{g_m} h \end{bmatrix} \cdot \frac{-\widehat{\mathcal{L}_f h} + \dot{y}_R + \alpha \left(y_R - y\right)}{\widehat{\mathcal{L}_g h}}.
\end{aligned}
$$

The proposed adaptive control formalism (36–37) can be efficiently applied wherever we have a problem of tracking a given signal with an output of a SISO–system (33–34) with unknown parameters.

4. Conclusion

In this paper we have presented two approaches to the human operator modeling: linear control theory approach and nonlinear control theory approach, based on the fixed and adaptive versions of a single-input single output Lie-Derivative controller. Our future work will focus on the generalization of the adaptive Lie-Derivative controller to MIMO systems. It would give us a rigorous closed–form model for model–free neural networks.

References

[1] Wickens, C.D., The Effects of Control Dynamics on Performance, in *Handbook of Perception and Human Performance,* Vol II, Cognitive Process and Performance (Ed. Boff, K.R., Kaufman, L., Thomas, J.P.), Wiley, New York, (1986).

[2] Wiener, N., *Cybernetics*, New York, (1961).

[3] Doherty, G., *Continuous Interaction and Manual Control, ERCIM News* **40**, January, (2000).

[4] Ivancevic, V., Ivancevic, T., *Geometrical Dynamics of Complex Systems.* Springer, Dordrecht, (2006).

[5] Ivancevic, V., Ivancevic, T., *Applied Differfential Geometry: A Modern Introduction.* World Scientific, Singapore, (2007).

[6] Phillips, J.M., Anderson, M.R., *A Variable Strategy Pilot Model, 2000 AIAA Atmospheric Flight Mechanics Conference,* Denver, Colorado, August, (2000).

[7] Phillips, J.M., *Variable Strategy Model of the Human Operator, PhD thesis in Aerospace Engineering,* Blacksburg, VI, (2000).

[8] McRuer, D.T., Jex, H.R., *A Review of Quasi-Linear Pilot Models, IEEE Transactions on Human Factors in Electronics*, Vol. HFE-**8**, 3, 231-249, (1967).

[9] McRuer, D.T., Krendel E.S., *Mathematical Models of Human Pilot Behavior, North Atlantic Treaty Organization Advisory Group for Aerospace Research and Development,* AGARD-AG-188, January, (1974).

[10] Kleinman, D.L., Baron, S., Levinson, W.H., *An Optimal Control Model of Human Response, Part I: Theory and Validation, Automatica,* **6**(3), 357-369, (1970).

[11] Sheridan, T.B., Ferrell, W., *Man-Machine Systems: Information, Control, and Decision Models of Human Performance,* MIT Press: Cambridge, MA, (1974).

[12] Innocenti, M., Belluchi, A., Balestrion, A., New Results on Human Operator Modelling During Non-Linear Behavior in the Control Loop, *1997 American Control Conference, Albuquerque, NM,* June 4-6 1997, Vol. 4, American Automatic Control Council, Evanston, IL, 2567-2570, (1997).

[13] Innocenti, M., Petretti, A., Vellutini, M., Human Operator Modelling During Discontinuous Performance, *1998 AIAA Atmospheric Flight Mechanics Conference*, Boston, Massachusetts, 31-38, August, (1998).

[14] McRuer, D., Allen, W., Weir, D., The Man/Machine Control Interface - Precognitive and Pursuit Control, *Proc. Joint Automatic Control Conference,* Vol. II, 81-88, Philadelphia, Pennsylvania, October, (1978).

[15] Young, L.R., Meiry, J.L., Bang-Bang Aspects of Manual Control in High-Order Systems, *IEEE Tr. Aut. Con. AC-***10**, 336-341, July, (1965).

[16] Pew, R.W., Performance of Human Operators in a Three-State Relay Control System with Velocity-Augmented Displays, *IEEE Tr. Human Factors, HFE-***7**(2), 77-83, (1966).

[17] Diamantides, N.D., A Pilot Analog for Aircraft Pitch Control, *J. Aeronaut. Sci.* **25**, 361-371, (1958).

[18] Costello, R.G., The Surge model of theWell-Trained Human Operator in SimpleManual Control, *IEEE Tr. Man-Mach. Sys. MMS-***9**(1), 2-9, (1968).

[19] Hess, R.A., A Rational for Human Operator Pulsive Control Behavior, *J. Guid. Con. Dyn.* **2**(3), 221-227, (1979).

[20] Phatak, A.V., Bekey, G.A., Model of the Adaptive Behavior of the Human Operator in Response to a Sudden Change in the Control Situation,
textit IEEE Tr. Man-Mach. Sys. MMS-**10**(3), 72-80, (1969).

[21] Pitkin, E.T., A Non-Linear Feedback Model for Tracking Studies, *Proceedings of the Eight Annual Conference on Manual Control,* University ofMichigan, Ann Arbor, Michigan, 11-22, May, (1972).

[22] Meritt, M.J., Bekey, G.A., An Asynchronous Pulse-Amplitude Pulse-Width Model of the Human Operator,
textit Third Annual NASA-University Conf. Manual Control, 225-239, March, (1967).

[23] Johannsen, G., Development and Optimization of a Nonlinear Multiparameter Human Operator Model, *IEEE Tr. Sys. Man, Cyber.* **2**(4), 494-504, (1972).

[24] Angel, E.S., Bekey, G.A., Adaptive Finite-State Models of Manual Control Systems, *IEEE Tr. Man-Mach. Sys.* **9**(1), 15-20, (1968).

[25] Costello, R., Higgins, R., An Inclusive Classified Bibliography Pertaining to Modeling the Human Operator as an Element in an Automatic Control System, IEEE Tr. Human Factors in Electronics, *HFE*-**7**(4), 174-181, (1966).

[26] Andrisani, D., Gau, C.F., A Nonlinear Pilot Model for Hover, *J. Guid. Con. Dyn.* **8**(3), 332-339, (1985).

[27] Heffley, R., Pilot Models for Discrete Maneuvers, *1982 AIAA Guidance, Navigation, and Control Conference Proceedings,* 132-142, San Diego, CA, August, (1982).

[28] Hess, R.A., Structural Model of the Adaptive Human Pilot, *J. Guid. Con. Dyn.* **3**(5), 416-423, (1980).

[29] Moorehouse, D., Modelling a Distracted Pilot for Flying Qualities Applications, *1995 AIAA Atmospheric Flight Mechanics Conference,* 14-24, Baltimore, Maryland, August, (1995).

[30] Ivancevic, V., Ivancevic, T., *Human-Like Biomechanics: A Unified Mathematical Approach to Human Biomechanics and Humanoid Robotics.* Springer, Dordrecht, (2005)

[31] Ivancevic, V., Ivancevic, T., *Complex Dynamics: Advanced System Dynamics in Complex Variables.* Springer, Dordrecht, (2007).

[32] Ivancevic, T., Jain, L. Pattison, J., Hariz, A., *Nonlinear Dynamics and Chaos Methods in Neurodynamics and Complex Data Analysis. Nonl. Dyn.* (Springer Online First)

[33] Ivancevic, V., Ivancevic, T., *High–Dimensional Chaotic and Attractor Systems.* Springer, Berlin, (2006).

[34] Ivancevic, V., Ivancevic, T., *Complex Nonlinearity: Chaos, Phase Transitions, Topology Change and Path Integrals, Springer, Series: Understanding Complex Systems,* Berlin, (2008).

[35] Ivancevic, V., Ivancevic, T., *Neuro-Fuzzy Associative Machinery for Comprehensive Brain and Cognition Modelling.* Springer, Berlin, (2007).

[36] Ivancevic, V., Ivancevic, T., *Computational Mind: A Complex Dynamics Perspective.* Springer, Berlin, (2007).

[37] Wilson, D., *Nonlinear Control, Advanced Control Course* (Student Version), Karlstad Univ. (2000).

[38] Spooner, J.T., Maggiore, M., Ordonez, R., Passino, K.M., *Stable Adaptive Control and Estimation for Nonlinear Systems: Neural and Fuzzy Approximator Techniques.* Wiley, New York, (2002).

[39] Brockett, R., *New Issues in the Mathematics of Control. In Mathematics Unlimited - 2001 and Beyond*, Springer, New York, (2001).

[40] Murray, R.M., Li, X., Sastry, S., *Robotic Manipulation,* CRC Press, Boco Raton, Fl, (1994).

[41] Isidori, A., *Nonlinear Control Systems. An Introduction,* (2nd ed) Springer, Berlin, (1989).

[42] Goodwine, J.W., *Control of Stratified Systems with Robotic Applications.* PhD thesis, California Institute of Technology, Pasadena, CA, (1998).

[43] Sastri, S.S., Isidori, A., Adaptive control of linearizable systems. *IEEE Tr. Aut. Ctrl.* **34**(11), 1123–1131, (1989).

[44] Nijmeijer, H., Van der Schaft, A.J., *Nonlinear Dynamical Control Systems.* Springer, New York, (1990).

[45] Gómez, J.C., Using symbolic computation for the computer aided design of nonlinear (adaptive) control systems. *Tech. Rep. EE9454*, Dept. Electr. and Comput. Eng., Univ. Newcastle, Callaghan, NSW, AUS, (1994).

[46] A.N. Michel, K. Wang, B. Hu, Qualitative Theory of Dynamical Systems (2nd ed.), Dekker, New York, (2001).

[47] Haken H., *Advanced Synergetics: Instability Hierarchies of Self–Organizing Systems and Devices (*3nd ed.). Springer, Berlin, (1993).

In: Computational Engineering
Editors: J. E. Browning and A. K. McMann
ISBN: 978-1-61122-806-9

Chapter 5

SOLUTION OF DISCRETE NONLINEAR EQUATION SYSTEMS RESULTING FROM THE FINITE ELEMENT METHOD USING A GLOBAL SECANT RELAXATION-BASED ACCELERATED ITERATION PROCEDURE

Chang-New Chen*
Department of Systems and Naval Mechatronic Engineering
National Cheng Kung University, Tainan, Taiwan

Abstract

A global secant relaxation(GSR)-based accelerated iteration scheme can be used to carry out the incremental/iterative solution of various nonlinear finite element problems. This computation procedure can overcome the possible deficiency of numerical instability caused by local failure existing in the iterative computation. Moreover, this method can efficiently accelerate the convergency of the iterative computation. This incremental/iterative analysis can consistently be carried out to update the response history up to a near ultimate load stage, which is important for investigating the global failure behaviour of a structure under certain external cause, if the constant stiffness is used. Consequently, this method can widely be used to solve general nonlinear problems.

Mathematical procedures of Newton-Raphson techniques in finite element methods for nonlinear finite element problems are summarized. These techniques are the Newton-Raphson method, quasi-Newton methods, modified Newton-Raphson methods and accelerated modified Newton-Raphson methods.

Numerical results obtained by using various accelerated modified Newton-Raphson methods are used to study the convergency performances of these techniques for material nonlinearity problems and deformation nonlinearity problems, separately.

Keywords: Discrete Nonlinear Equation Systems; Finite Element Method; Differential Quadrature Finite Element Method; Incremental/Iterative Solution Procedure; Global Secant Relaxation; Accelerated Iteration Procedure; Newton-Raphson Method; Quasi-Newton Methods; Modified Newton-Raphson Methods

*E-mail address: cchen@mail.ncku.edu.tw

1. Introduction

An efficient and reliable incremental/iterative solution procedure, based on using a global secant relaxation technique to adjust the constant stiffness prediction and improve the convergency, for nonlinear finite element problems which have geometric and/or material nonlinearities has been proposed by the author (Chen, 1990,1992). The accelerated incremental/iterative solution procedure has also been used to accelerate the standard modified Newton-Raphson methods (Chen, 1993,1994). A predictor-corrector procedure for the iterative solution of linear or nonlinear finite element systems using the diagonal stiffness matrix or constant diagonal stiffness matrix has also been proposed (Chen, 1991,1995).

Efficient and reliable solution of a static nonlinear finite element system with equilibrium iteration is an important topic in the area of scientific and engineering computation since advanced design is a challengeable trend in current and future technology. The global secant relaxation-based accelerated modified Newton-Raphson iteration techniques can be incorporated into the static nonlinear finite element solution procedure to construct an efficient and reliable algorithm.

In this chapter, the numerical procedures for solving static nonlinear finite element problems by using the incremental procedure and the global secant relaxation-based accelerated modified Newton-Raphson iteration techniques are described. It has been proved that the algorithm is efficient and reliable. It is also thought that the accuracy is high due to the adoption of equilibrium iteration. Consequently, the nonlinear problems can be efficiently and reliably solved by this accelerated incremental/iterative numerical procedure.

In order to further reduce the computational cost, the differential quadrature finite element method (DQFEM) (Chen, 1998) proposed by the author is used to carry out the finite element discretization. This approach can reduce the arithmetic operations in calculating the static or dynamic incremental equilibrium equations.

The method of DQ defines a set of nodes in a problem domain. Then a derivative or partial derivative of a variable function at a node with respect to a coordinate is approximated as a weighted linear sum of all the function values at all nodes along that coordinate direction (Bellman and Casti, 1971). The DQ has been generalized which leads to the generic differential quadrature (GDQ) (Chen, 1999,2006). The weighting coefficients for a grid model defined by a coordinate system having arbitrary dimension can also be generated. The configuration of a grid model can be arbitrary. In the GDQ, a certain order derivative or partial derivative of the variable function with respect to the coordinate variables at a node is expressed as the weighted linear sum of the values of function and/or its possible derivatives or partial derivatives at all nodes.

The DQ and GDQ have been extended which results in the extended differential quadrature (EDQ) (Chen, 2000,2006). In the EDQ discretization, a discretization can be defined at a point which is not a node. The points for defining the discretizations are discrete points. In using the EDQ to FEM, the integration points are typical discrete points. A node can also be a discrete point. Then a certain order derivative or partial derivative of the variable function with respect to the coordinate variables at an arbitrary discrete point can be expressed as the weighted linear sum of the values of function and/or its possible derivatives or partial derivatives at all nodes. By using DQ, GDQ and EDQ, the DQFEM has been developed.

Numerical results of sample nonlinear problems solved by using GSR-based accelerated modified Newton-Raphson iteration, constant stiffness iteration and constant diagonal stiffness iteration, separated, are included to show the efficiency and reliability of these accelerated incremental/iterative solution techniques.

2. DQFEM Discretization

2.1. FEM Discretization

For a deformable body under elastic-plastic deformation condition, let Δu_i denote the incremental displacement vector, the incremental form of Green-Saint Venant strain is defined as follows:

$$\Delta\varepsilon_{lm} = \frac{1}{2}(\Delta u_{l,m} + \Delta u_{m,l} + \Delta u_l \Delta u_m) \tag{1}$$

For elastic-plastic analysis, the incremental strain $\Delta\varepsilon_{lm}$ at some stage after initial yielding can be separated into elastic component $\Delta\varepsilon_{lm}^e$ and plastic component $\Delta\varepsilon_{lm}^p$

$$\Delta\varepsilon_{lm} = \Delta\varepsilon_{lm}^e + \Delta\varepsilon_{lm}^p \tag{2}$$

Introducing the elastic constitutive matrix D_{kjlm} into the generalized Hooke's law, the incremental stress $\Delta\sigma_{kj}$ can be obtained

$$\Delta\sigma_{kj} = D_{kjlm}(\Delta\varepsilon_{lm} - \Delta\varepsilon_{lm}^p) \tag{3}$$

From the theory of plasticity, it is known that the yield surface can be described by the following function relation:

$$F(\sigma_{kj}, \varepsilon_{kj}^p, \kappa) = 0. \tag{4}$$

in which κ is a strain hardening parameter. And by adopting F as the plastic potential and considering the normality rule, the increment of plastic strain during plastic deformation can be expressed as

$$\Delta\varepsilon_{kj}^p = \Delta\lambda A_{kj} \tag{5}$$

where $\Delta\lambda$ is a scalar parameter, and

$$A_{kj} = \frac{\partial F}{\partial\sigma_{kj}} \tag{6}$$

Using (3) and (4), the following elastic-plastic constitutive relation can be obtained (Zienkiewicz, 1977)

$$\Delta\sigma_{kj} = D_{kjlm}^{ep}\Delta\varepsilon_{lm} \tag{7}$$

where

$$D_{kjlm}^{ep} = (D_{kjlm} - \frac{D_{kjrs}A_{rs}A_{pq}D_{pqlm}}{A_{rs}A_{pq}D_{rspq} + b})$$

$$b = -\frac{1}{\Delta\lambda}(\frac{\partial F}{\partial\varepsilon_{kj}^p}\Delta\varepsilon_{kj}^p + \frac{\partial F}{\partial\kappa}\Delta\kappa), \quad \Delta\lambda = \frac{A_{kj}D_{kjlm}\Delta\varepsilon_{lm}}{A_{kj}D_{kjlm}A_{lm} + b}$$

For isotropic hardening materials described by work hardening hypothesis, b is equivalent to the strain hardening rate H'.

The total Lagrangian formulation is used for carrying out the finite element discretization. For load stage o, let $u_i^{o,n}(x_h)$, $\varepsilon_{lm}^{o,n}(x_h)$ and $\sigma_{kj}^{o,n}(x_h)$ denote the updated displacement strain and stress of iteration step n, respectively, in a finite element domain Ω^e. Also let $\Delta u_i^{o,n+1}(x_h)$, $\Delta\varepsilon_{lm}^{o,n+1}(x_h)$ and $\Delta\sigma_{kj}^{o,n+1}(x_h)$ denote the incremental displacement strain and stress of iteration step $n+1$, respectively. Then the updated displacement $u_i^{o,n+1}(x_h)$, strain $\varepsilon_{lm}^{o,n+1}(x_h)$ and stress $\sigma_{kj}^{o,n+1}(x_h)$ of iteration step $n+1$ can be expressed as the following form:

$$\begin{aligned} u_i^{o,n+1}(x_h) &= u_i^{o,n}(x_h) + \Delta u_i^{o,n+1}(x_h) \\ \varepsilon_{lm}^{o,n+1}(x_h) &= \varepsilon_{lm}^{o,n}(x_h) + \Delta\varepsilon_{lm}^{o,n+1}(x_h) \\ \sigma_{kj}^{o,n+1}(x_h) &= \sigma_{kj}^{o,n}(x_h) + \Delta\sigma_{kj}^{o,n+1}(x_h) \end{aligned}$$

Let f_i^o denote the body force of load stage o. The equilibrium condition after updating the incremental information of iteration step $n+1$ can be expressed as the following equation:

$$[\sigma_{kj}^{o,n+1}(\delta_{ij} + u_{i,j}^{o,n+1})]_{,k} = -f_i^o \tag{8}$$

In the above equation, the updated $u_{i,j}^{o,n}$ and $\sigma_{kj}^{o,n}$ are necessary for defining $u_{i,j}^{o,n+1}$ and $\sigma_{kj}^{o,n+1}$. Also let n_k and T_i^{o+1} denote the outward unit normal vector and traction force, respectively, on the element boundary. The traction condition is expressed by

$$T_i^{o,n+1} = n_k\sigma_{kj}^{o,n+1}(\delta_{ij} + u_{i,j}^{o,n+1})$$

Using the interpolation function, denoted as $\Psi_\alpha(x_h)$, the incremental and updated displacements in an element can be discretized as follows:

$$\begin{aligned} \Delta u_i^{o,n+1}(x_h) &= \Psi_\beta(x_h)\Delta u_{i\beta}^{o,n+1} \\ u_i^{o,n+1}(x_h) &= \Psi_\beta(x_h)u_{i\beta}^{o,n+1} \end{aligned} \tag{9}$$

A Galerkin procedure can be constructed by using the equilibrium condition of (8) and the interpolation function $\Psi_\alpha(x_h)$ which results in obtaining the following discretized equation:

$$\begin{aligned} &([\sigma_{kj}^{o,n+1}(\delta_{ij} + u_{i,j}^{o,n+1})]_{,k}\, , \Psi_\alpha) = (-f_i^o\, , \Psi_\alpha) \\ &- < (T_i^{o,n+1} - n_k\sigma_{kj}^{o,n+1}(\delta_{ij} + u_{i,j}^{o,n+1}))\, , \Psi_\alpha > \end{aligned} \tag{10}$$

In the above equation $(\, .\, ,\, .\,)$ and $<\, .\, ,\, .\, >$ represent the integrations over the element and element boundary, respectively. Integrating by parts for (10) and considering the traction condition, then introducing (1), (6), (7) and (9) into the resulting equation the equilibrium equation for iteration step $n+1$ of load stage o for an element can be obtained which is expressed as the following form:

$$k_{i\alpha h\beta}^{o,n}\Delta u_{h\beta}^{o,n+1} = r_{i\alpha}^{o,n} \tag{11}$$

where

$$\begin{aligned} k_{i\alpha h\beta}^{o,n} &= (\sigma_{kj}^{o,n}\Psi_{\beta,j}\delta_{ih}\, , \Psi_{\alpha,k}) \\ &+(D_{kjlm}^{ep}(\delta_{ij} + u_{i,j}^{o,n})(\delta_{lh} + u_{h,l}^{o,n})\Psi_{\beta,m}\, , \Psi_{\alpha,k}) \end{aligned} \tag{12}$$

is the tangent stiffness matrix updated at the end of iteration step n, and

$$r_{i\alpha}^{o,n} = (f_i^o\,, \Psi_\alpha) + < T_i^{n+1}\,, \Psi_\alpha > + < T_i^{n+1}\,, \Psi_\alpha > - (\sigma_{kj}^{o,n}(\delta_{ij} + u_{i,j}^{o,n})\,, \Psi_{\alpha,k}) \tag{13}$$

a force vector resulted from subtracting the internal force vector updated at end of iteration step n from the force vector due to body force and traction force. It should be mentioned that the concept of weak formulation is considered when assemble the incremental element stiffness equations into the overall incremental stiffness equation. It should also be mentioned that the residual force vector can be expressed by the incremental response quantities and external forces.

For the elastic-plastic analysis a proper returning scheme must be used to set the final stress state on the yield surface if the resulting stress state defined by the elastic trial stress lies outside of the elastic region enclosed by the yield surface.

3. EDQ Discretization

The EDQ is used to discretize the derivatives or partial derivatives of the displacements with respect to the coordinate variables at integration points or certain other points interested in carrying out the FEM formulation. Since the displacements can be expressed by using the interpolation functions the derivatives or partial derivatives of the interpolation functions with respect to the coordinate variables can be expressed by the weighting coefficients. Thus the EDQ weighting coefficients for one element in an element group can also be used to discretize all other elements in that element group. It can reduce the CPU time required for calculating the discrete dynamic equilibrium equations.

Consider the two-dimensional elements with the displacements being approximated by using certain analytical functions of two coordinate variables. The dimensions for defining the discrete point and node can be different. By adopting a one-dimensional node identification method to express both the discrete point and node, the EDQ discretization for a $(m+n)$th order partial derivative of the displacements u_i at the discrete point L can be expressed by

$$\frac{\partial^{(m+n)} u_{iL}}{\partial x^m \partial y^n} = D_{Lr}^{x^m y^n}\, \tilde{u}_{ir}, \quad r = 1, 2, ..., N_D \tag{14}$$

where N_D is the number of degrees of freedom and $\tilde{u}_{iL}$ the values of displacements and/or its there possible partial derivatives at the N_N nodes. Each of the displacements can be a set of appropriate analytical functions denoted by $\Upsilon_p(x, y)$. The substitution of $\Upsilon_p(x, y)$ in (14) for any displacement leads to a linear algebraic system for determining $D_{Lr}^{x^m y^n}$. The set of analytical functions can also be expressed by a tensor having an order other than one. The displacements can also be approximated by

$$u_i(x, y) = \Psi_p(x, y)\tilde{u}_{ip}, \quad p = 1, 2, ..., N_D \tag{15}$$

where $\tilde{u}_{ip}$ are the values of displacements and/or there possible partial derivatives at the N_N nodes, and $\Psi_p(x, y)$ their corresponding interpolation functions. Adopting the set of $\Psi_p(x, y)$ as each displacement $u_i(x, y)$, the same procedure can also be used to find $D_{Lr}^{x^m y^n}$.

And the $(m+n)$th order partial differentiation of (15) at discrete point L also leads to the EDQ discretization equation (14) in which $D_{Lr}^{x^m y^n}$ is expressed by

$$D_{Lr}^{x^m y^n} = \frac{\partial^{(m+n)} \Psi_r}{\partial x^m \partial y^n} |_L \tag{16}$$

The displacements can also be approximated by

$$u_i(x,y) = \Upsilon_p(x,y) c_{ip}, \quad p = 1, 2, ..., N_D \tag{17}$$

The constraint conditions at all discrete points can be expressed as

$$\tilde{u}_{iq} = \chi_{qp} c_{ip} \tag{18}$$

where χ_{qp} are formed by the values of $\Upsilon_p(x,y)$ and their possible partial derivatives at all nodes. Using (17) and (18), the weighting coefficients can also be obtained

$$D_{Lr}^{x^m y^n} = \frac{\partial^{(m+n)} \Upsilon_{\bar{q}}}{\partial x^m \partial y^n} |_L \chi_{r\bar{q}}^{-1}$$

In (17), the unknown coefficients and appropriate analytical functions can also be expressed by certain other tensors having orders other than one.

It should be mentioned that the coordinate variables x and y can be either physical or natural. It should also be mentioned that the appropriate analytical functions can be formed by using certain basis functions defined by the coordinate variables independently. The basis functions can be the polynomials, sinc functions, Lagrange interpolated polynomials, Chebyshev polynomials, Bernoulli polynomials, Hermite interpolated polynomials, Euler polynomials, rational functions, ..., etc. To solve problems having singularity properties, certain singular functions can be used. The problems having infinite domains can also be treated.

4. Equilibrium Iteration

A generic nonlinear equation system can be solved by the nonlinear iteration. The generalized nonlinear iteration techniques such as the nonlinear Jacobi, Gauss-Seidel, successive over-relaxation(SOR), Peaceman-Rachford iterations, ..., ect. are typical iteration methods.

The generalized-linearized methods such as the Newton, secant and Steffensen iterations are simplified nonlinear iterative methods. The nonlinear iteration can be carried out by combining a generalized-linearized method with a certain linear iterative method. By adopting a linear iterative procedure as the primary iteration and a generalized-linearized method as the secondary iteration, in the nonlinear iteration, it results in a linear-nonlinear iteration scheme. The Jacobi-, Gauss-Seidel-, SOR- and Peaceman-Rachford-, etc. Newton, secant and Steffensen iterations are typical linear-nonlinear iteration schemes.

By reversing the roles of the linear iterative method and the generalized-linearized method it leads to the composite nonlinear-linear iteration procedure with the generalized-linearized method as the primary iteration and the linear iterative method as the secondary iteration. The Newton-, secant- and Steffesen-, etc. Jacobi, Gauss-Seidel, SOR, and

Peaceman-Rachford iterations are typical nonlinear-linear iteration schemes. The composite nonlinear-linear iteration can be generalized to solve the generalized-linearized system not only by the linear iterations but also by certain other solvers such as the direct methods. The quasi-Newton methods, modified Newton-Raphson methods and accelerated modified Newton-Raphson methods are also generalized-linearized iteration schemes.

Certain popular FEM nonlinear equilibrium iteration techniques are summarized. The mathematical procedures of numerical algorithms are described.

4.1. The Newton-Raphson Method

In using the Newton-Raphson method to the equilibrium iteration, the stiffness matrix K_{rs} and residual force vector have to be updated for each iteration step. The stiffness matrix can thus be constructed. Then the incremental stiffness equation for the iteration step $n+1$ of a specific load stage o can be expressed by

$$K_{rs}^{o,n} \Delta U_s^{o,n+1} = R_r^{o,n} \tag{19}$$

where $K_{rs}^{o,n}$ is the stiffness matrix, $\Delta U_s^{o,n+1}$ the incremental displacement vector and $R_r^{o,n}$ the residual force vector. By solving (19) to obtain $\Delta U_s^{o,n+1}$, the updated displacement vector of iteration step $n+1$ can be obtained

$$U_s^{o,n+1} = U_s^{o,n} + \Delta U_s^{o,n+1} \tag{20}$$

4.2. Quasi-Newton Methods

Quasi-Newton methods use finite difference approximations to the derivative operations to update a secant stiffness matrix at each iteration step. The information of two consecutive iteration steps is necessary in order to construct a secant stiffness matrix. It can overcome the possible difficulty caused by the exact evaluation of derivative operations in updating a tangent stiffness matrix to obtain a stiffness matrix for the Newton-Raphson iteration.

Broyden's method is a generalization technique of the secant method. Let $\bar{K}_{rs}^{o,n}$ denote the secant stiffness matrix of this method. $\bar{K}_{rs}^{o,n}$ can be updated by using the following equation

$$\bar{K}_{rs}^{o,n} = \bar{K}_{rs}^{o,n-1} + \frac{\left[(R_r^{o,n-1} - R_r^{o,n}) - \bar{K}_{rt}^{o,n-1} \Delta U_t^{o,n}\right] \Delta U_s^{o,n}}{\Delta U_t^{o,n} \Delta U_t^{o,n}} \tag{21}$$

Using $\bar{K}_{rs}^{o,n}$ to replace $K_{rs}^{o,n}$ in (19), the incremental displacement vector $\Delta U_s^{o,n+1}$ can be predicted by this quasi-Newton method.

Using Sherman and Morrison's matrix inversion formula, $\left(\bar{K}_{rs}^{o,n}\right)^{-1}$ can be directly updated without decomposing $\bar{K}_{rs}^{o,n}$. This can improve the numerical efficiency significantly. Consider a nonsingular matrix A_{rs} and two vectors p_r and q_s, the inverse of $A_{rs} + p_r q_s$ is

$$\left(A_{rs} + p_r q_s\right)^{-1} = A_{rs}^{-1} - \frac{A_{rl}^{-1} p_l q_m A_{ms}^{-1}}{1 + q_l A_{lm}^{-1} p_m} \tag{22}$$

Letting $A_{rs} = \bar{K}_{rs}^{o,n-1}$, $p_r = (R_r^{o,n-1} - R_r^{o,n}) - \bar{K}_{rt}^{o,n-1} \Delta U_t^{o,n}$ and $q_s = \Delta U_s^{o,n}$, (21) together with (22) implies that

$$
\begin{aligned}
(\bar{K}_{rs}^{o,n})^{-1} &= \left(\bar{K}_{rs}^{o,n-1} + \frac{[(R_r^{o,n-1} - R_r^{o,n}) - \bar{K}_{rt}^{o,n-1}\Delta U_t^{o,n}]\Delta U_s^{o,n}}{\Delta U_t^{o,n}\Delta U_t^{o,n}}\right)^{-1} \\
&= (\bar{K}_{rs}^{o,n-1})^{-1} \\
&\quad - \frac{(\bar{K}_{rl}^{o,n-1})^{-1}\{[(R_l^{o,n-1} - R_l^{o,n}) - \bar{K}_{lt}^{o,n-1}\Delta U_t^{o,n}]\Delta U_m^{o,n}/\Delta U_t^{o,n}\Delta U_t^{o,n}\}(\bar{K}_{ms}^{o,n-1})^{-1}}{1 + \Delta U_l^{o,n}(\bar{K}_{lm}^{o,n-1})^{-1}[(R_m^{o,n-1} - R_m^{o,n}) - \bar{K}_{mt}^{o,n-1}\Delta U_t^{o,n}]/\Delta U_t^{o,n}\Delta U_t^{o,n}} \\
&= (\bar{K}_{rs}^{o,n-1})^{-1} - \frac{[(\bar{K}_{rl}^{o,n-1})^{-1}(R_l^{o,n-1} - R_l^{o,n}) - \Delta U_r^{o,n}]\Delta U_m^{o,n}(\bar{K}_{ms}^{o,n-1})^{-1}}{\Delta U_l^{o,n}(\bar{K}_{lm}^{o,n-1})^{-1}(R_m^{o,n-1} - R_m^{o,n})} \\
&= (\bar{K}_{rs}^{o,n-1})^{-1} + \frac{[\Delta U_r^{o,n} - (\bar{K}_{rl}^{o,n-1})^{-1}(R_l^{o,n-1} - R_l^{o,n})]\Delta U_m^{o,n}(\bar{K}_{rs}^{o,n-1})^{-1}}{\Delta U_l^{o,n}(\bar{K}_{lm}^{o,n-1})^{-1}(R_m^{o,n-1} - R_m^{o,n})}
\end{aligned} \tag{23}
$$

In using this quasi-Newton method to the equilibrium iteration, the stiffness matrix of the first iteration step, for each time stage, has to be updated. The stiffness matrix can thus be formed and decomposed. Then (23) is used to update the inverse matrices of all subsequent iteration steps. This updating procedure involves only matrix multiplication at each iteration step. The condition for $(\bar{K}_{rs}^{o,n})^{-1}$ to be singular is that $\Delta U_l^{o,n}$ and $(R_m^{o,n-1} - R_m^{o,n})$ are orthogonal relative to $(\bar{K}_{lm}^{o,n-1})^{-1}$.

4.3. Modified Newton-Raphson Methods

In using the standard modified Newton-Raphson scheme to the equilibrium iteration, only the stiffness matrix of the first iteration step, for each time stage, is necessary to be updated. The stiffness matrix can thus be constructed. Letting K_{rs}^o denote the stiffness matrix of the first step of time stage o. Then the incremental stiffness equation can be expressed by

$$
K_{rs}^o \Delta U_s^{o,n+1} = R_r^{o,n} \tag{24}
$$

In applying (24) to the equilibrium iteration, K_{rs}^o can also be replaced by a stiffness matrix formed at a certain other iteration response stage in the incremental/iterative integration solution.

4.4. Accelerated Modified Newton-Raphson Methods

The convergence of modified Newton-Raphson methods can be improved by using certain procedures to adjust the incremental response vector obtained by the modified Newton-Raphson prediction at each iteration step. The accelerated modified Newton-Raphson schemes using a global secant relaxation (GSR) procedure, proposed by the author (Chen, 1990), are described.

Through the introduction of an implicit secant stiffness matrix, denoted as $\tilde{K}_{rs}^{o,n}$, the incremental displacement vector obtained by solving (24) can be used to construct a secant relation for hardening or softening response behaviours. This secant relation is shown to have the following form:

$$
\tilde{R}_r^{o,n} = R_r^{o,n} \pm \tilde{K}_{rs}^{o,n} \Delta U_s^{o,n+1} \tag{25}
$$

in which $\tilde{R}_r^{o,n}$ is a residual force vector after the modified Newton-Raphson prediction. The GSR method uses an accelerator defined by minimizing certainly defined system error to scale the incremental displacement vector. Let $\omega^{o,n+1}$ denote this accelerator, the updated displacement vector after acceleration can be expressed as

$$U_s^{o,n+1} = U_s^{o,n} + \omega^{o,n+1} \Delta U_s^{o,n+1} \tag{26}$$

and a residual force vector after acceleration can be expressed by the following form:

$$\bar{R}_r^{o,n+1} = R_r^{o,n} \pm \omega^{o,n+1} \tilde{K}_{rs}^{o,n} \Delta U_s^{o,n+1} \tag{27}$$

This residual force vector provides important information for defining the system error.

In defining the system error, consistency and reliability have to be considered, in the sense that the defined error must be able to better reflect the true error existing in the system. The Euclidean norm of $\bar{R}_r^{o,n+1}$ is a good quantity for representing the system error. This quantity can be expressed as

$$E_f^{o,n+1} = \bar{R}_r^{o,n+1} \bar{R}_r^{o,n+1} \tag{28}$$

By using (25) and (27) in (28), then minimizing $E_f^{o,n+1}$ with respect to $\omega^{o,n+1}$, an accelerator denoted as $\omega_f^{o,n+1}$ can be obtained which shows to have the following form:

$$\omega_f^{o,n+1} = \mp \frac{R_r^{o,n}(R_r^{o,n} - \tilde{R}_r^{o,n})}{(R_s^{o,n} - \tilde{R}_s^{o,n})(R_s^{o,n} - \tilde{R}_s^{o,n})} \tag{29}$$

The related acceleration scheme is GSR-MR. By using a diagonal matrix with all diagonal elements having the same value to replace the implicit secant stiffness matrix, (29) will result in the formula representing the Generalized Aitken accelerator proposed by Cahill (1992).

It is worth mentioning that no mathematical approximation is involved in deriving the accelerator $\omega_f^{o,n+1}$. Therefore, in considering the exact description of the secant relation of (25) and the direction of the incremental response vector, the resulting iterative procedure is believed to be a highly consistent secant improvement-based iteration scheme.

The energy norm defined as the inner product of residual force vector $\bar{R}_r^{o,n+1}$ and the incremental displacement vector caused by $\bar{R}_r^{o,n+1}$, along the linear deformation surface is also used to evaluate the system error. This error can be expressed as

$$E_e^{o,n+1} = \bar{R}_r^{o,n+1} (\tilde{K}_{rs}^{o,n})^{-1} \bar{R}_s^{o,n+1} \tag{30}$$

By using the same procedures as those used in defining $\omega_f^{o,n+1}$, another accelerator denoted as $\omega_e^{o,n+1}$ can be obtained:

$$\omega_e^{o,n+1} = \mp \frac{\Delta U_r^{o,n+1} R_r^{o,n}}{\Delta U_s^{o,n+1}(R_s^{o,n} - \tilde{R}_s^{o,n})} \tag{31}$$

The improvement scheme using $\omega_e^{o,n+1}$ as the accelerator is GSR-MW.

It should be noted that the accelerator $\omega_e^{o,n+1}$ is shown to have the same form as the single-parameter accelerator proposed by Crisfield (1984) though the fundamental concepts of acceleration, the mathematical formulations and the resulting iterative algorithms are actually different. In investigating the formulation procedures of these two accelerators, it is believed that this similarity is caused by the fact that both of these two approaches use the secant relation to approximately describe the deformation behaviours of the two states used to construct the secant relation.

It is also valuable to investigate the difference of convergency performances between GSR-MR and GSR-MW theoretically. As already mentioned previously, the mathematical formulation of defining $\omega_f^{o,n+1}$ is absolutely consistent. Therefore, the resulting iterative scheme is believed to be highly reliable and the convergency performance should be good. On the other hand, inconsistency does exist in the formulation procedure in defining $\omega_e^{o,n+1}$ since a mathematical approximation of using $\tilde{K}_{rs}^{o,n}$ to predict the incremental response vector is used to construct the energy norm-based system error. This approximation will result in obtaining a less reliable evaluation of the system error, which will lead to a less reliable acceleration scheme with less promising convergency performance. And the inconsistency will be even severe for solving nonlinear finite element problems if large time increments are used.

The accelerated constant stiffness iteration is one of the accelerated modified Newton-Raphson methods. This scheme uses the linear elastic stiffness matrix to predict the initial incremental displacement vector for all incremental/iterative steps.

The linear equation systems existing in the generalized-linearized DQFEM nonlinear iterations can be solved by using a certain direct or iterative solver. The most commonly used direct solvers are Gauss elimination, Cholesky decomposition and frontal method. Various techniques including the sparse implementation strategies, the domain decomposition and the parallel implementation was considered in implementing an efficient direct solution procedure into a DQFEM computer program. There are also many iterative solvers that can be used to solve a linear equation system. Among the indirect solvers the preconditioned conjugate gradient (PCG) methods have been attracting lots of the finite element programmers. In solving large linear equation systems, the PCG methods can offer promising performances due to the substantial reductions in computer memory requirements and the function of taking the advantage of vector and parallel processing strategies on computers that support these features. The iterative solvers possess a relatively high degree of natural concurrency, with the predominant operations in PCG algorithms being saxpy operations, inner products and matrix-vector multiplications. Among the PCG algorithms, the stabilized and accelerated version of the biconjugate gradient method, which is an extension of the conjugate gradient method to nonsymmetric systems, is one of the most commonly used iterative solvers. The element-by-element solution procedure is also a useful algorithm which has considerable operation count and I/O advantages since no overall stiffness matrix is needed to be formed.

4.5. Explicit Predictor-Corrector Iteration

The author has also proposed a diagonal stiffness-based predictor-corrector procedure for iteratively solving linear or nonlinear finite element equation systems (Chen, 1995). It is an explicit iteration procedure in the nonlinear iteration. Instead of using an assembled overall stiffness matrix, this method only uses the diagonal elements of the overall stiffness matrix to predict the incremental displacement vector in carrying out the iterative solution.

Consequently, only the diagonal elements of the element stiffness matrices are needed to be calculated. Thus the computer memory requirement can be minimized. Let $K_s^{o,n}$ denote the vector representing the set of the diagonal elements in the stiffness matrix $K_{rs}^{o,n+1}$. Then, by using $K_s^{o,n}$ and referring (19) the following equation can be constructed.

$$K_{(s)}^{o,n} \Delta U_s^{o,n+1} = R_s^{o,n} \tag{32}$$

Equations (37) and (31) represent the explicit predictor-corrector equilibrium iteration procedure. The discrete equation system can also be solved without updating the vector representing the set of the diagonal stiffness elements for each iteration step. Then, by modifying (29) the following equation can be constructed

$$K_{(s)}^{o} \Delta U_s^{o,n+1} = R_s^{o,n} \tag{33}$$

Equations (33) and (26) represent a different version of the explicit predictor-corrector equilibrium iteration procedure.

For solving a linear equation system such as the incremental equation of the step-by-step procedure, which might adopt the unbalance load correction instead of adopting the equilibrium iteration, the overall stiffness can be formed and used to calculate the residual force vectors required for the predictor-corrector iteration. However, it needs to save the overall stiffness matrix in the computer memory unit. Let ΔF_s^o denote the incremental load vector. The incremental equation of the step-by-step procedure for load stage o is expressed by

$$K_{rs}^{o} \Delta U_s^{o} = \Delta F_r^{o} \tag{34}$$

Then the residual force vectors used to define the scaling factor can be calculated by

$$R_r^{o,n} = \Delta F_r^{o} - K_{rs}^{o} \sum_{k=1}^{n} \Delta U_s^{o,k} \tag{35}$$

$$\tilde{R}_r^{o,n} = R_r^{o,n} - K_{rs}^{o} \Delta U_s^{o,n+1} \tag{36}$$

The predictor-corrector iterative procedure needs less computer storage space. It is also suitable for vector and parallel implementation. In applying this iterative procedure to solve a generic equation system, the amplification of longer period errors can be prevented by the introduction of a GSR correction which contributes to greater numerical stability. Thus all longer and shorter period errors can be effectively eliminated by this predictor-corrector solution. Numerical results have proved that this iterative procedure has good numerical stability. It is also an efficient algorithm. This iterative procedure can also be used to the multi-grid solution in which the longer period errors, which may not be efficiently eliminated by the fine grid iteration, can be substantially reduced by the use of a coarse grid correction. The coarse grid can adopt either iterative or direct procedure.

In implementing the DQFEM analysis program, various phases including preprocessing, calculation of elemental discrete equations, incorporation of boundary conditions, solution of system equations and postprocessing can be parallelized. However, the assembly of elemental discrete equations can not take the advantage of parallel operation efficiently.

5. Sample Solutions

5.1. GSR-Based Accelerated Constant Stiffness Iteration

A problem involving the static elastic-plastic analysis of a square plate with a square cutout, subjected to uniformly distributed axial load, was solved. The deformation is assumed to be linear. The description of the model problem can be seen in Fig. 1 in which the thickness of the plate is $1mm$. Elastic-perfectly plastic material with Young's modulus E being equal to $21000kg/mm^2$, yield stress $\sigma_Y = 36.3kg/mm^2$ and Poisson's ratio $\nu = 0.3$, was considered. The modelling considers the symmetry property. Eight-node quadratic element adopting 3×3 quadrature rule was used to discretize one quarter of the plate domain. The mesh is shown in Fig. 2. The accelerated constant stiffness iteration was used to efficiently and reliably carry out the iterative computation in which Tresca's yield criterion was used to detect the plastification of integration points. The constant stiffness matrix was formed at the beginning of the incremental/iterative solution. The convergence indicator is defined as the ratio of residual norm to the norm of load vector, which was selected to be 10^{-6}. The response history was updated up to a near collapse load stage under which the structure showed globally unstable behaviour due to the entire loss of elastic restoring capability caused by the formation of an unstable failure mechanism. The load-displacement curve with the displacement at point A for the plate model with $a/W = 0.5$ is shown in Fig. 3 in which the predicted value of the collapse load is $0.38\sigma_Y$. Convergency tests were carried out by using a single larger load increment of $0.3\sigma_Y$ to perform the equilibrium iteration. The convergence indicator is defined as the ratio of residual norm to the norm of load vector. The convergency curves of three different iteration algorithms were plotted. They can be seen in Fig. 4. It shows that the GSR acceleration has good numerical stability and convergency performance though the load increment is large. It also shows that GSR-MR performs the best. The collapse load analysis was also carried out for the model problems with $a/W = 0.15$, 0.3 and 0.8, respectively. The predicted collapse loads for the four model analyses were plotted in Fig. 5. They were compared with the results of Hodge's bound solutions (Hodge, 1959). It shows that the results of DQFEM solutions tend to be close to the upper bound of the results obtained by classical limit strength analysis, which reflects the engineering reality inherent in the DQFEM discretization of using displacement model.

5.2. GSR-Based Accelerated Modified Newton-Raphson Iteration

A problem involving the elastically nonlinear analysis of a clamped circular plate subjected to a uniformly distributed lateral force, was solved. The model problem is shown in Fig. 6 in which ten axially symmetric 8-node isoparametric elements possessing quadratic deformation behaviour are used to discretize the plate domain. A response history curve was updated and shown in Fig. 7 in which the adequacy of finite element discretization was ensured by the fact that numerical results were in good agreement with theoretical solutions. The indicator of convergence, denoted as Δ^n and defined as the logarithm of the ratio of residual norm to the norm of the vector of boundary nodal forces, was selected to be -4. A process of using four load increments to update the response history curve was used to study the convergency performances of three different iteration schemes. The numbers of

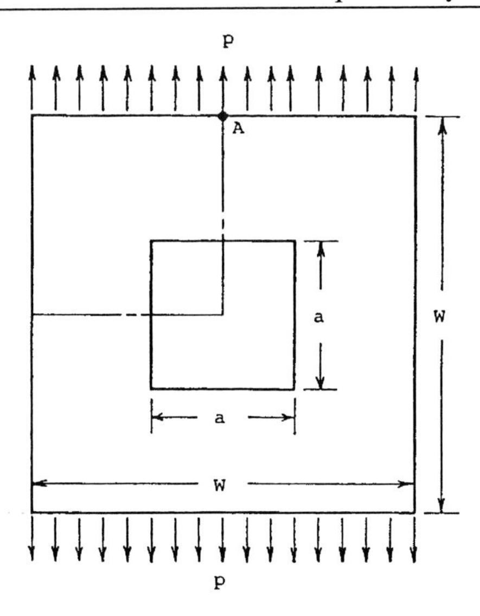

Figure 1. A square plate with a square cutout.

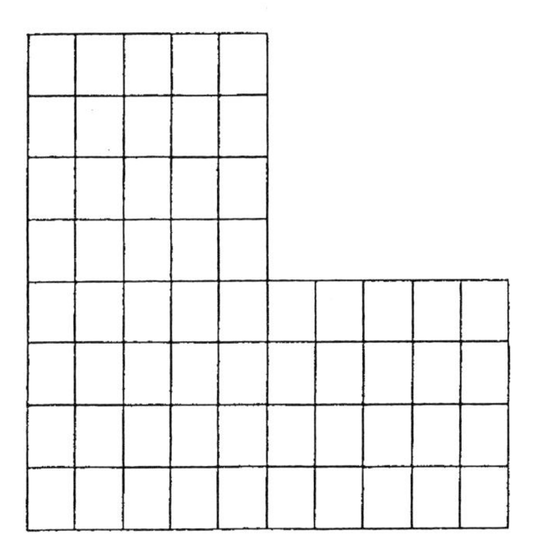

Figure 2. The mesh.

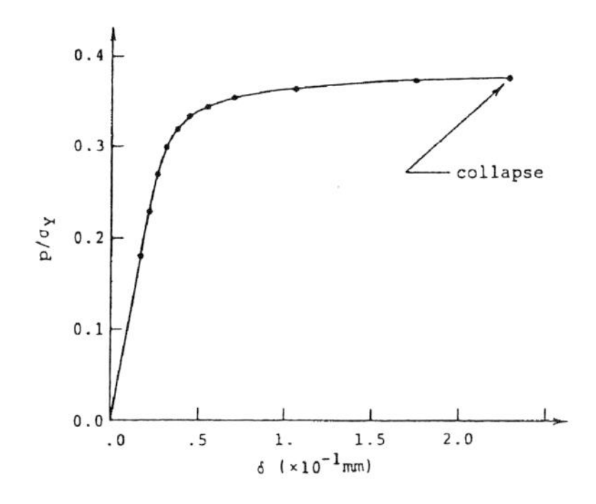

Figure 3. Load-displacement curve of point A for $a/W = 0.5$.

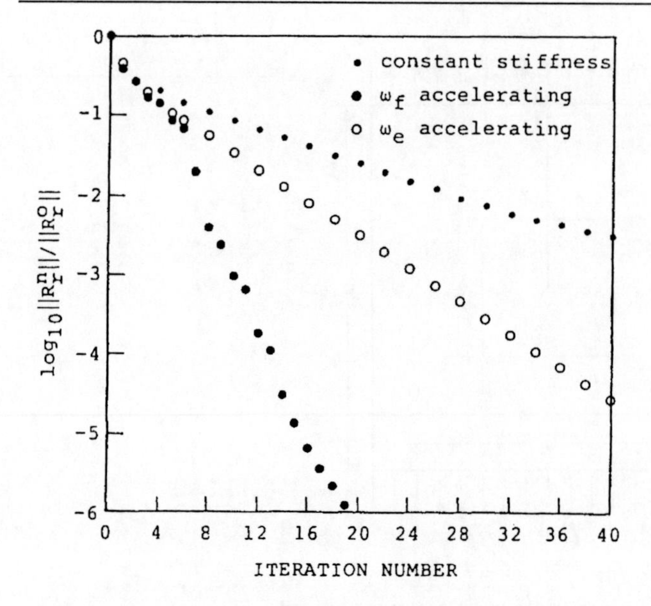

Figure 4. The convergence of equilibrium iteration. (load increment $\Delta p^1 = 0.3\sigma_Y$).

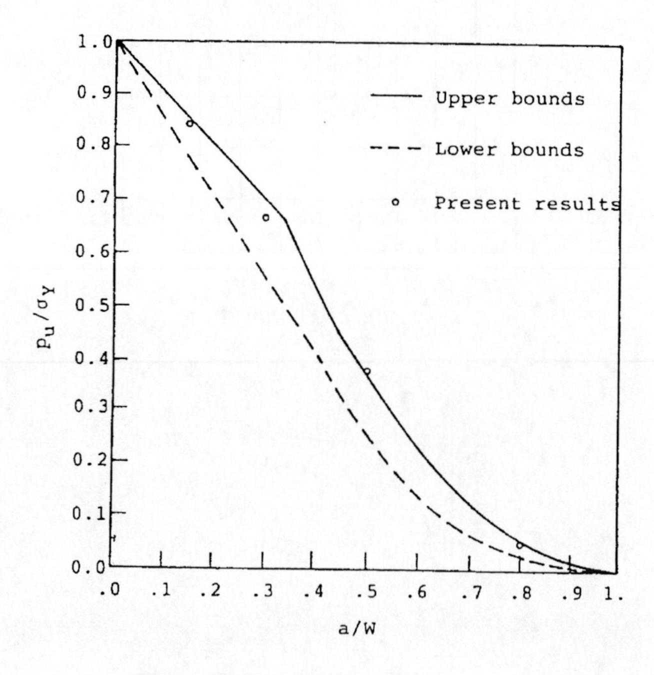

Figure 5. Ultimate collapse loads of the square plate with a square cutout.

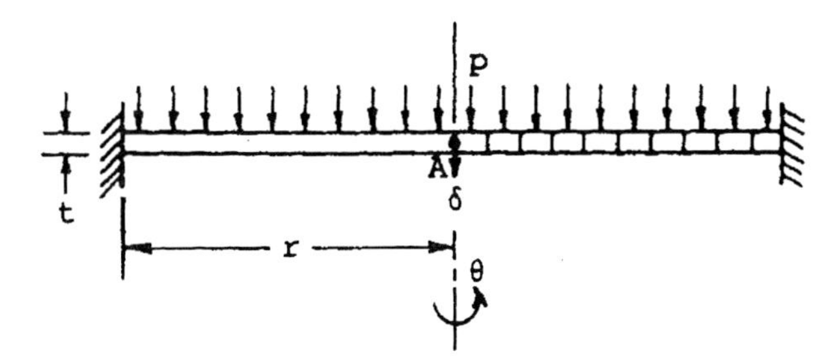

Figure 6. A clamped circular plate.

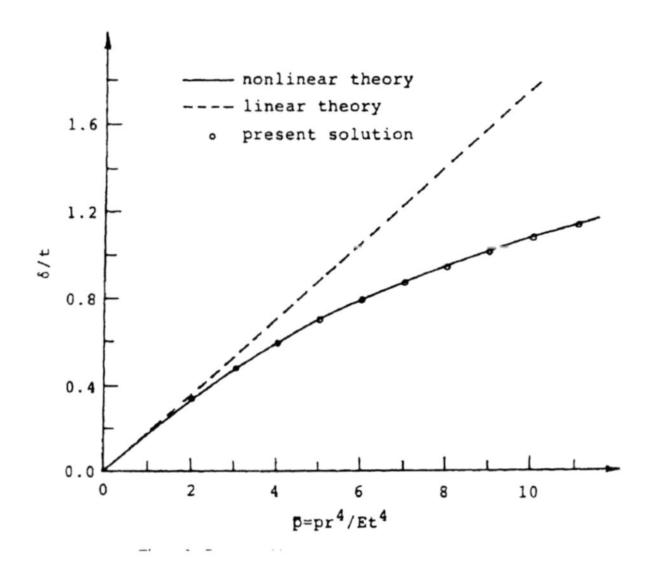

Figure 7. Response history curve of circular plate analysis.

iterations required for the solutions to converge are listed in Table 1 with which it is known that the GSR improvement accelerates the convergence of the modified Newton-Raphson iteration significantly. It also shows that GSR-MR performs slightly better than GSR-MW.

5.3. GSR-Based Accelerated Constant Diagonal Stiffness Iteration

The iteration procedure can be used to solve either linear or nonlinear problems. A problem solved involves the geometrically nonlinear two-dimensional elasticity structure. Dimensionless parameters were used to describe the geometry, material properties of plates, and external forces applied. A square plate, shown in Fig. 8, with side length and thickness being equal to 8. and 1., respectively, was considered. The Young's modulus of the material is 7000. and the Poisson's ratio is .33. Only concentrated point loads were applied to the structure. The arrow symbol is used to represent a load unit applied at the specified point in the arrow direction. The bilinear element was used to model the square plate and the ratio of residual norm to the norm of load vector was selected to be the indicator of convergence. Three analyses using pure diagonal stiffness iteration, GSR-MR-based and

Table 1. Iteration numbers required for convergence of circular plate analysis

Accelerator $\bar{p} = pr^4/Et^4$	None	ω_f	ω_e
2	242	12	14
4	34	8	10
7	39	8	8
10	11	6	6
Total	326	34	38

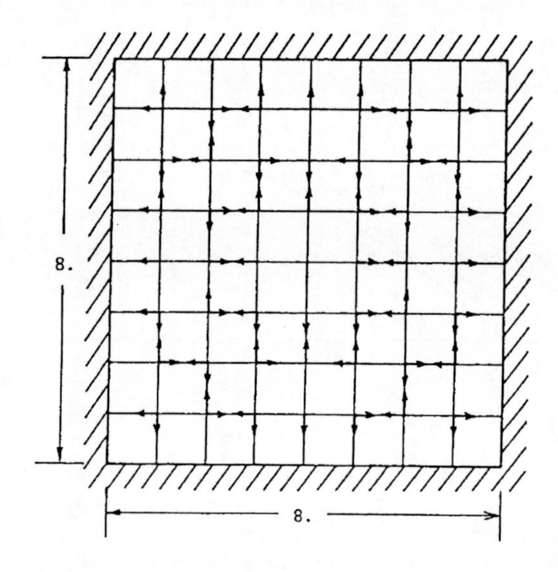

Figure 8. A fixed plate subjected to concentrated inplane forces.

GSR-MW-based predictor-corrector iterations were carried out.

A certain high frequency loading condition was applied which could be seen in Fig. 8. In order to know the influence of degree of nonlinearity to the convergence behaviour of this model problem, three analyses using different load units were carried out. The resulting convergence information were plotted in Figs. 9, 10 and 11. From these figures, it is known that for the GSR-based predictor-corrector iteration method no clear tendency of the influence of degree of nonlinearity to the convergence rate can be seen. However, the convergence of pure diagonal stiffness iteration is largely dependent on the degree of nonlinearity of the nonlinear system analyzed. It also shows that the predictor-corrector iteration method has excellent numerical stability and convergence performance. Moreover, it shows that the GSR-MR corrector always performs the best.

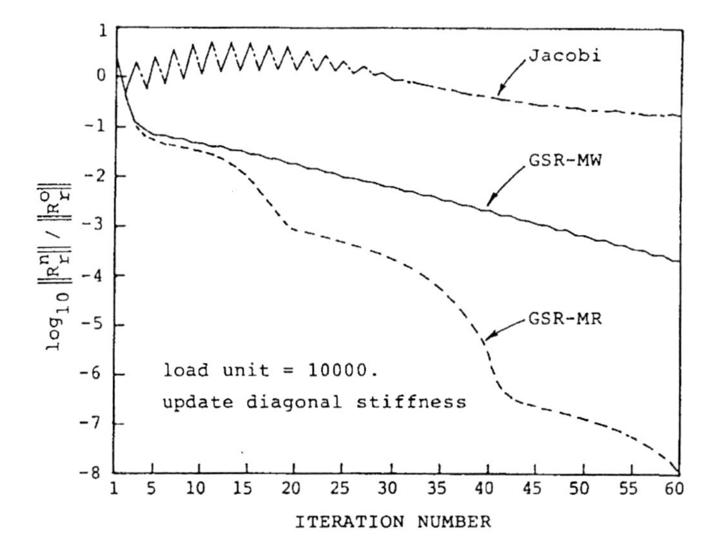

Figure 9. Convergence of load 1.

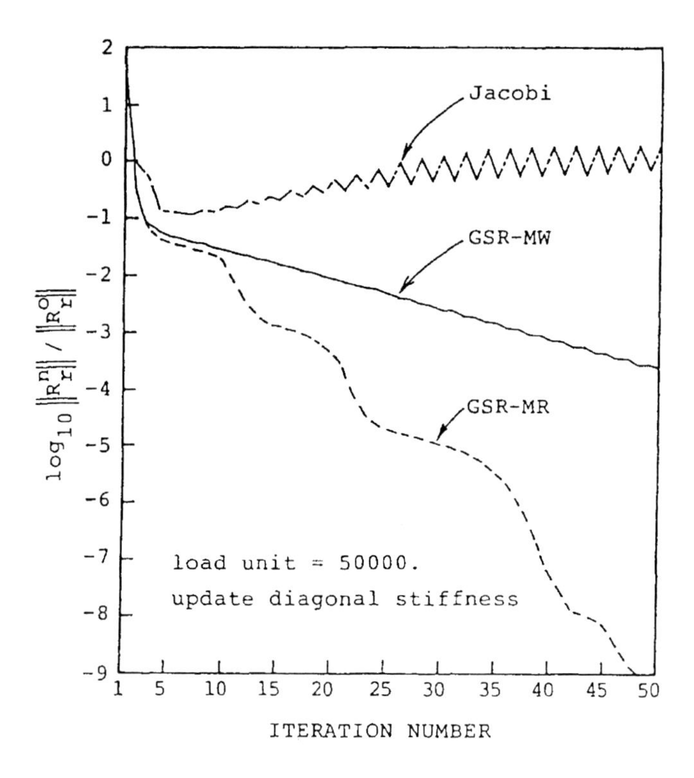

Figure 10. Convergence of load 2.

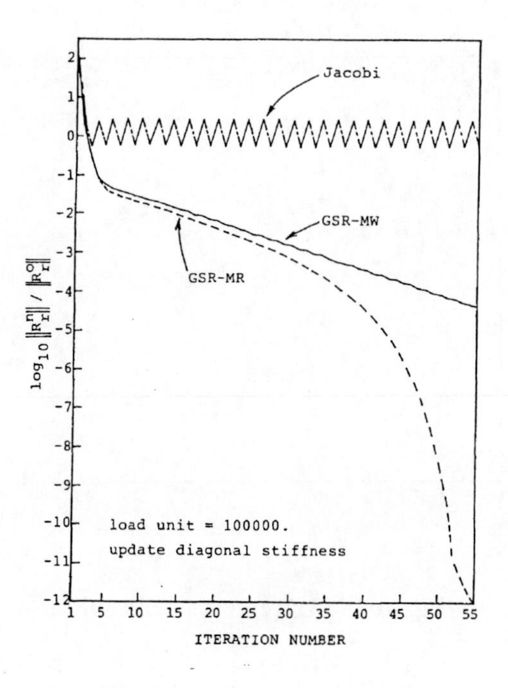

Figure 11. Convergence of load 3.

6. Conclusions

The global secant relaxation(GSR)-based accelerated iteration schemes were summarized and presented. Numerical results of some sample problems solved using various accelerated modified Newton-Raphson methods proved that these schemes cab be used to carry out the incremental/iterative solution of various nonlinear finite element problems, efficiently and reliably.

References

Bellman, R.E. and Casti, J., 1971, "Differential Quadrature and Long-term Integration," *J. Math. Anal. Appl.,* Vol. 34, pp. 235-238.

Cahill, E., 1992, "Acceleration Techniques for Functional Iteration of nonlinear Equations", *IMACS Conference on Mathematical Modelling and Scientific Computing,* Bangalove, INDIA.

Chen, C.N., 1990, "Improved Constant Stiffness Algorithms for the Finite Element Analysis", *Proc. NUMETA* **90**, pp. 623-628, Swansea, UK.

Chen, C.N., 1991, "A Predictor-corrector Procedure for the Iterative Solution of Algebraic Systems Resulting from the Discretization of Continuum Problems", *First US Natl. Congr. Comput. Mechs.*, Chicago, USA.

Chen, C.N., 1992, "Efficient and Reliable Accelerated Constant Stiffness Algorithms for The Solution of nonlinear Problems", *Intl. J. Numer. Methods Engr.*, Vol. 35, No., 481-490.

Chen, C.N., 1993, "An Accelerated Modified Newton-Raphson Procedure as Applied to The Analysis of Geometrically Nonlinear Finite Element Problems", American Academy of Mechanics, *Proc. 3rd Pan Amer. Congr. Appl. Mechs.,* 89-92, Sao Paulo, BRAZIL.

Chen, C.N., 1994, "Accelerated Modified Newton-Raphson Iteration-based Finite Element Solution of Structural Problems Involving Nonlinear Deformation Behavior", *Commun. Numer. Methods Engr.*, Vol. 10, No. 4, 333-338.

Chen, C.N., 1995, "A Global Secant Relaxation (GSR) Method-Based Predictor-Corrector Procedure for the Iterative Solution of Finite Element Systems", *Comput. Struct.,* Vol. 54, pp. 199-205.

Chen, C.N., 1999, "Generalization of Differential Quadrature Discretization", *Numerical Algorithms,* Vol. 22, No. 2, 167 182.

Chen, C.N., 2000, "Differential Quadrature Element Analysis Using Extended Differential Quadrature", *Comput. Math. Appls.,* Vol. 39, No. 5-6, 65-79.

Chen, C.N., 2006, Discrete Element Analysis Methods of Generic Differential Quadratures - Volume 25, *Series of Lecture Notes in Applied Computational Mechanics,* Springer, Berlin, GERMANY.

Crisfield, M.A., 1984, "Accelerated and Damping the Modified Newton-Raphson Method", *Comput. Struct.*, Vol. 18, pp. 267-278.

Hodge, P.G., 1959, "Plastic Analysis of Structures", McGraw-Hill, New York.

Zienkiewicz, O.C., 1977, *The Finite Element Method,* McGraw-Hill, New York.

INDEX

D

E

F

G

H

I

J

K

L

M

N

O

P

Q

R

S

T

U

V

W

Y